HEALTH AND SAFETY IN WELDING
AND ALLIED PROCESSES

Third edition

Edited by N.C.Balchin, MA, PhD

THE WELDING INSTITUTE
Abington Hall Abington Cambridge CB1 6AL

First edition March 1956, Institute of Welding
Revised and enlarged July 1963
Second edition, edited by P.J.R.Challen, March 1965
Third edition 1983, The Welding Institute

Although every care has been taken, the
Institute cannot be held responsible for the
consequences of any errors in or omissions from
this new edition

© 1983

ISBN 0 85300146 4

CONTENTS

			Page
Introduction to the Third Edition			vii
Chapter	1	**PROCESS FUNDAMENTALS** Introduction — The gas flame — The electric arc — Other arc processes — Electrical resistance — Other heat sources	1
	2	**GAS CYLINDERS AND GAS WELDING AND CUTTING EQUIPMENT** High pressure plant	17
	3	**LOW PRESSURE PLANT**	28
	4	**SAFETY PRECAUTIONS DURING GAS WELDING AND CUTTING OPERATIONS** Explosion risks — Fire risks — Eye injuries and burns — Fume risks — Lighting	32
	5	**PROTECTIVE CLOTHING, AND EYE AND HEAD PROTECTION, FOR GAS WELDING AND CUTTING** Protective clothing for the body — Protection for the eyes and head	36
	6	**THE CARE OF ARC WELDING AND CUTTING EQUIPMENT** Equipment connected to an electricity supply — Engine-driven equipment — All equipment and welding circuits — Electrode holders — Accessories — Gas-shielded arc welding — Metal-arc gas-shielded welding — Tungsten-arc gas-shielded welding	40
	7	**SAFETY PRECAUTIONS DURING ARC WELDING AND CUTTING OPERATIONS** Electric shock — Burns — Eye injuries — Fume risks — Fire risks	46
	8	**PROTECTIVE CLOTHING, AND EYE AND HEAD PROTECTION, FOR ARC WELDING** Protective clothing for the body — Protection for the	53

eyes and head — Arc cutting — Ear protection in arc cutting and other processes

9	**PRECAUTIONS FOR WELDING AND CUTTING VESSELS WHICH HAVE HELD COMBUSTIBLES** Small vessels — Large vessels — Permit to work	60
10	**PLASMA ARC PROCESSES** Plasma arc — Physical hazards — Toxic hazards — Protective equipment	65
11	**ELECTRO-SLAG WELDING AND CONSUMABLE-GUIDE WELDING** Toxic hazards — Physical hazards	68
12	**RESISTANCE WELDING** Electrical hazards — Physical hazards — Toxic hazards — Noise	69
13	**THE THERMIT PROCESS** Safety precautions — Ventilation — Protective equipment	72
14	**ELECTRON-BEAM WELDING** Electrical hazards — Radiation hazards — Toxic hazards — Physical hazards	76
15	**LASER WELDING AND CUTTING** Electrical hazards — Radiation hazards	81
16	**BRAZING AND BRAZE WELDING** Brazing processes — Explosion hazards — Toxic hazards — Surface preparation and cleaning procedures — Furnace atmospheres — Electrical hazards — Physical hazards — Braze welding	84
17	**SOFT SOLDERING** Processes — Toxic hazards — Electrical hazards — Physical hazards — Fire hazards	93
18	**THERMAL SPRAYING** Surface preparation by blasting — Toxic hazards — Physical and electrical hazards — Explosion and fire hazards — Thermal spraying equipment — Recovery of overspray	97
19	**WELDING AND FLAME SPRAYING PLASTICS** Welding plastics — Toxic hazards — Flame spraying plastics	103

| 20 | RADIOGRAPHIC INSPECTION | 108 |

Special precautions

| 21 | MECHANICAL HAZARDS | 112 |

Safe platforms — Obstructions — Lifting — Manipulators and positioners — Wire feed units — Testing — Cutting — Grinding — Protection — Guarding machinery

| 22 | MEASUREMENT AND ASSESSMENT OF FUME | 115 |

Measurement of fume concentrations — Threshold Limit Values

| 23 | SOURCES OF FUME | 118 |

Surface treatment — Macroetching — Action of the heat source on the parent metal — Action of the heat source on the surface coating of the parent metal — Action of the heat source on the surrounding air — Action of the heat on the consumables — Internal combustion engines

| 24 | VENTILATION AND FUME PROTECTION | 133 |

Working conditions — Ventilation — Personal fume protection — Recommended practice

| 25 | LIGHTING | 143 |

Normal lighting — Emergency lighting

| 26 | FIRE | 145 |

Fire prevention — Detection — Rescue — Preventing the spread of fire — Extinguishing fires

| 27 | FIRST AID | 153 |

Introduction — Electric shock and burns — Eye injuries — Major wounds — Minor cuts and abrasions — Burns of the skin — Traumatic shock — Heat stroke — Exposure to harmful gases and fumes — First aid equipment

| 28 | LEGISLATION ON WELDING SAFETY IN THE UK | 161 |

Introduction — The Health and Safety at Work etc. Act, 1974 — The Fire Precautions Act, 1971 — The Factories Act, 1961 — The Electricity Regulations, 1908 — The Blasting (Castings and Other Articles) Special Regulations, 1949 — The First Aid Boxes in Factories Order, 1959 — The First Aid (Standard of Training) Order, 1960 — The Shipbuilding and

Ship-repairing Regulations, 1960 — The Construction (General Provisions) Regulations, 1961 — The Ionising Radiations (Sealed Sources) Regulations, 1969 — The Abrasive Wheels Regulations, 1970 — The Highly Flammable Liquids and Liquefied Petroleum Gases Regulations, 1972 — The Protection of Eyes Regulations, 1974 — The Explosives Act, 1875 — Orders in Council (No.30) Prohibiting the Manufacture, Importation, Keeping, Conveyance, or Sale of Acetylene when an Explosive as defined by the Order, 1937 — The Petroleum (Consolidation) Act, 1928 — The Petroleum (Carbide of Calcium) Order, 1929 — The Petroleum (Compressed Gases) Order, 1930 — The Gas Cylinder (Conveyance) Regulations, 1931 — Employment Protection Act, 1974 — Civil proceedings

APPENDIX 1	Bibliography of further information	168
APPENDIX 2	Useful addresses	181
APPENDIX 3	Metric units	186
APPENDIX 4	Updating	189
INDEX		193

INTRODUCTION TO THE THIRD EDITION

Since the second edition of this handbook was published in 1965 a number of developments have contributed to the need for this fully revised edition. In particular there has been a change of emphasis: instead of traditional working practices being abandoned or modified only if they are shown to be unsafe, the trend is now to require positive proof that work can be carried on safely. This has been confirmed in the United Kingdom by the introduction of the Health and Safety at Work Act, 1974.

This handbook has been compiled on the basis that its main use will be in the UK and, where no specific qualification is made, information as to legislation, colour codes, etc. is intended to refer to the UK only. Most of the information will, however, be of much wider application if allowance is made for variations such as those of law and climate.

The contents of this edition are similar to those of the second but have been revised and extended to take into account modern developments both in welding processes and in research on their hazards. Note that the safety of welded structures does not fall within the scope of this book.

The first two editions were published under the imprint of the Institute of Welding which, in 1968, amalgamated with the British Welding Research Association to form The Welding Institute. Apart from the general programme of the Institute's research laboratories on the development of welding processes of all kinds, which must necessarily include their safe application to industry, for several years there has also been a specific investigation of a number of aspects of airborne pollutants. It is therefore appropriate that this edition appears in an extensive list of Welding Institute publications.

The information in this book has been checked and revised with the assistance of a panel of experts having much experience of the application of welding in industry, and it is believed to be correct and accurate as far as normal welding, cutting, and allied operations are concerned. However, it is not possible to cover every eventuality in a book of this size, and further advice should always be sought if a situation arises which presents special hazards.

Chapter 1 defines and explains welding and allied processes very briefly: the previous editions have been found of use not only to welding staff, but to a much wider range of people who are responsible for safety without necessarily having any specialised knowledge of welding, such as safety officers, line supervision including foremen and managers, and industrial medical staff. Chapters 2 to 5 cover gas welding processes, and 6 to 8 arc welding. Chapter 9 states the precautions which are necessary when welding and cutting vessels contaminated by combustibles. Chapters 10-18 cover the more specialised welding processes: brazing, soldering, and thermal spraying, and Chapter 19 the welding and flame spraying of plastics. Chapter 20 deals with radiographic inspection. The remaining chapters cover problems common to most processes: mechanical hazards, flame and ventilation, lighting, fire, and first aid, concluding with Chapter 28 on UK legislation.

This book has been kept reasonably short for ease of reference and, in the many instances where it has not been possible to include complete information, an indication has been given as to where this is reasonably accessible. This has been done by listing printed and other material in Appendix 1, and useful organisations in Appendix 2. The book concludes with Appendix 3, a brief list of relevant metric units with appropriate conversions, and Appendix 4, updating information to January 1982.

The objective of this book is to identify the hazards involved in welding and to show how they should be controlled to provide safe working conditions for the welder and other personnel. In this way the Institute hopes that it will make an important contribution to health and safety in industry. Suggestions relating to the content of the next edition will be welcomed.

N.C. Balchin

CHAPTER 1

PROCESS FUNDAMENTALS

INTRODUCTION

This chapter presents very briefly the fundamental principles of the various processes discussed in the book so that the origin of the hazards may be understood. The processes are classified according to the heat source employed. The terminology generally follows that given in British Standard 499 'Welding terms and symbols', but common usage is also indicated where it differs from the Standard; some of the terms used are Trade Marks or trade names which should be applied to products or services of their registered proprietors only.

A number of heat sources may be used for more than one process and the industrial processes based on each heat source are shown in italics below the descriptions; the following general definitions apply:

Welding: making a union between pieces of metal or thermoplastic at joint faces rendered plastic or liquid by heat or by pressure, or both

Brazing: making a joint between pieces of metal in which molten filler metal is drawn by capillary attraction into the space between the closely adjacent surfaces of the parts to be joined. In general, the melting point of the filler metal is above 500°C and below that of the metal to be joined

Braze welding: the making of a joint between pieces of metal using a filler metal with a lower melting point than that of the sections to be joined. Capillary action does not enter into the process

Soldering: a joining process similar to brazing but in which the filler metal melts at a temperature below 500°C

Surfacing: the deposition of a layer of a metal on a substrate of the same or different metal by a process involving heat

Thermal spraying: spraying material which has been rendered molten in a spray gun. The molten material is in a finely divided form and is projected on to the surface to be coated by compressed air. This surface has been suitably prepared to receive the sprayed material.

Thermal cutting: parting or shaping materials by the application of heat with or without a stream of cutting oxygen

The three most important direct sources of heat are the:

(a) flame produced by the combustion of a fuel gas with air or oxygen
(b) electric arc struck between an electrode and the workpiece, or between two electrodes
(c) electrical resistance offered to the passage of a current between two or more workpieces

THE GAS FLAME

Supplies of oxygen or air and a fuel gas are fed to a blowpipe where they are mixed prior to combustion at a nozzle. The flame produced by this combustion consists of two zones: the inner cone-shaped zone at the extremity of which the highest temperature is found, and the outer zone surrounding the cone which is at a lower temperature and which shields any hot or molten metal from the atmosphere. The inner zone can be set to a neutral, oxidising, or excess fuel gas condition, and all three flame conditions have various uses.

Gas welding; braze welding; torch brazing; torch soldering; heating

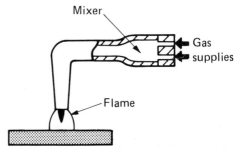

Compressed air/fuel gas or entrained air/fuel gas flames burn at temperatures lower than those produced with oxygen:

Fuel gas	Flame temperature, °C	
	With oxygen	With air
Acetylene	3260	2325
Propane	3000	1950
Coal gas	2200	1850

Gases are usually supplied from cylinders at a high pressure which is reduced to that required by a regulator. (Some older equipment supplied one gas at a high pressure and the other at a low pressure, Chapter 3. The high pressure gas, after regulation, draws up and carries along the other gas.)

The metal is heated to ignition temperature by a flame, and a jet of pure oxygen is directed on to the heated surface at the point of cutting. An oxidation reaction occurs progressively as the nozzle is moved along the line to be cut. When cutting metals which have refractory oxides, such as stainless steels or aluminium, a powder may be fed to the cutting zone. This may be either:

(a) iron powder, which burns in the oxygen, and dissolves the oxide (obsolescent), or
(b) abrasive, cutting away the oxide

Oxygen cutting; gas cutting; oxygas cutting; flame cutting; gouging and lancing; powder cutting

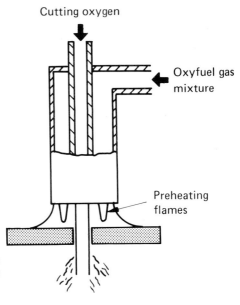

THE ELECTRIC ARC

The workpieces and an electrode are connected to a power supply (AC or DC) and an arc is struck between them; its temperature is normally over 4000°C. The workpieces melt and are fused together. If it is necessary to add extra ('filler') metal to the joint this may be achieved by either melting the electrode itself (consumable electrode) or using an electrode which does not melt (nonconsumable electrode) and melting a separate filler rod or wire which is not carrying current.

Nonconsumable electrode processes

In these processes, transfer of arc energy raises the temperature of the workpiece to melting point; for welding, this is employed to melt both the edges of the components to be joined and the filler rod. In some processes, the arc is struck between two electrodes.

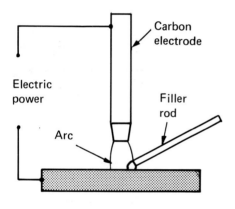

The arc is struck between a carbon electrode and the workpiece.

Carbon arc (obsolescent)

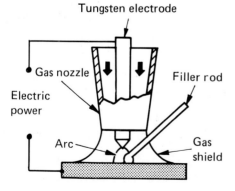

A shield of inert gas is used to protect the electrode, filler, and weld metal from attack by the atmosphere. The electrode is of tungsten and the gas is argon, helium, nitrogen, or mixtures of these gases with small quantities of hydrogen. The arc may be started and maintained by a high voltage high frequency supply superimposed on the main welding current.

Tungsten-inert-gas, TIG; tungsten-arc gas shielded; argon arc; heliarc; heliweld; gas tungsten arc welding, GTAW.

Consumable electrode processes

In these processes an arc is struck between a consumable electrode and the workpiece. The arc energy raises the temperature of both the workpiece and the end of the electrode to melting point. Metal melts off the electrode and is transferred across the arc to the weld. The electrode is advanced to maintain a constant arc length.

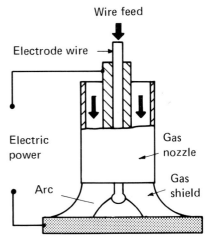

The arc and weld are shielded by a gas, either carbon dioxide, argon, helium, or a mixture of these, with or without small additions of oxygen. The solid wire electrode is fed in by a motor to maintain a constant arc length.

CO_2; metal-inert-gas, MIG; metal-active-gas, MAG; metal-arc gas-shielded; semi-automatic; gas-metal arc welding, GMAW.

A variant of the above uses a hollow wire with a flux core which generates gas to assist shielding, with or without further gas from the nozzle.

Flux-cored; inner shield; self-shielded

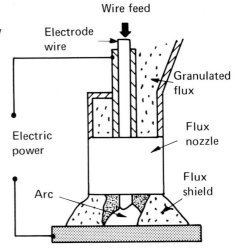

The arc is submerged beneath a covering of granulated flux which protects the arc zone and the weld from atmospheric attack and takes part in metallurgical reactions with the molten weld metal. The electrode is a bare wire which is automatically fed to maintain a constant arc length.

Submerged-arc welding

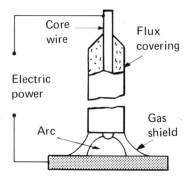

The electrode is covered with a flux, which reacts during welding to form a shielding gas to protect the arc zone and the molten weld pool, and a slag to protect the cooling weld metal. The slag also takes part in metallurgical reactions with the molten weld metal. Current is fed into the electrode at the far end, usually via a hand-held electrode holder.

Manual metal-arc, MMA; electric arc; stick welding; gravity welding; touch welding; shielded metal-arc welding, SMAW.

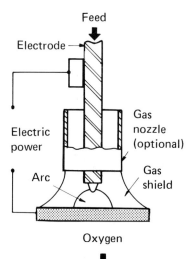

A flexible electrode is made by wrapping wire spirals round a core wire and filling the gaps with flux; this is fed mechanically into the arc.

Automatic metal-arc; Fusarc

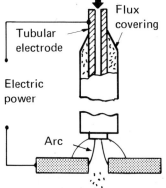

An arc is struck between a tubular covered electrode and the workpiece; oxygen is injected into the tubular core. The energy of the arc raises the workpiece to oxidation temperature and an oxidation reaction occurs progressively as the arc is moved along the line to be cut.

Oxygen arc cutting; oxyarc cutting

An arc is struck between the end of a carbon electrode and the workpiece. A pool of molten metal is produced which is then blown away by jets of compressed air. The process of melting and ejection proceeds continuously as the arc moves along the line to be cut or gouged.

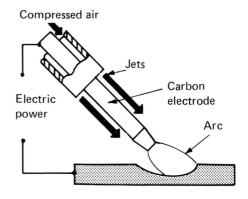

Air-arc cutting

Other arc processes

An arc is struck between the component to be attached and the structure. The arc energy raises the temperature of the end of the component and, locally, the structure to melting point. The component is then automatically pressed into the molten weld pool and a weld is effected.

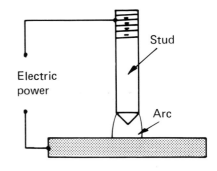

Stud welding; arc stud welding

The arc is constricted by passing it through a relatively small orifice through which a stream of gas is also flowing. The gas may be air, argon, helium, nitrogen, hydrogen, or mixtures of these gases. The arc plasma, consisting of ionised gas at a temperature of about 24 000°C, is formed into a jet by the gas pressure and continues as a 'flame' beyond the nozzle.

Plasma welding; plasma cutting; microplasma welding; plasma spraying; constricted-arc; needle-arc

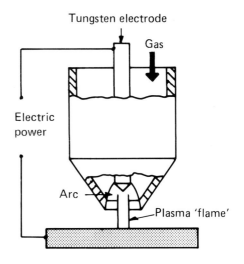

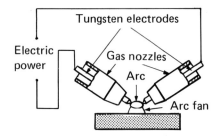

An arc is struck between two tungsten electrodes and a jet of hydrogen is directed into it. Dissociation of hydrogen into atomic form is followed by recombination in a characteristic fan-shaped flame with the evolution of heat; it is this heat which is used for welding.

Atomic hydrogen welding (obsolescent)

ELECTRICAL RESISTANCE

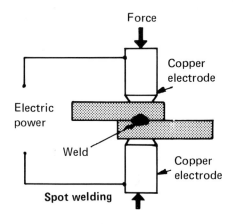

A high current at a low voltage flows through the two components from the electrodes. The electrodes are of very high conductivity material so that most of the heat is generated by the resistance offered to the current at the interface between the components, thus bringing the contiguous areas of the components (the 'faying' surfaces) to welding temperature. During passage of the current, and for some time afterwards, a force is applied by the electrodes to forge the weld.

Spot welding; seam welding; projection welding; butt welding; electric resistance welding (ERW)

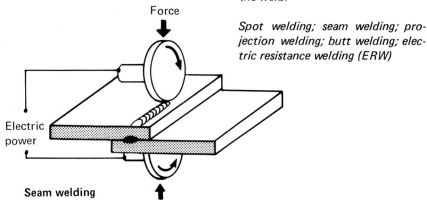

When the ends of the components are brought into contact, 'flashing' occurs accompanied by the violent expulsion of molten particles. This brings the ends of the components to welding temperature and a weld is obtained by the application of a forging force.

Flash welding

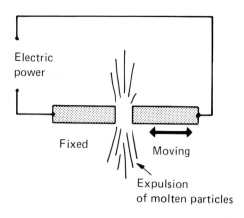

The workpieces are set in a vertical plane with a gap between them and copper plates are placed on the sides of the joint. An arc is established under flux between one or more continuously fed electrode wires and a metal plate at the bottom of the joint. A pool of molten metal is formed. When molten, the flux or slag becomes an electrical conductor. The current is now carried from the electrode to the workpieces through the slag. The heat melts the sides of the joint and the electrode wire. As welding progresses, the copper plates or shoes are moved up.

Electro-slag welding

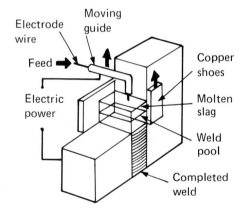

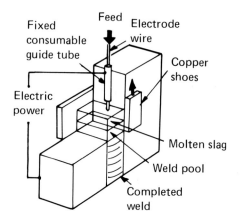

In a variant of the above, a fixed consumable guide tube positioned in the joint melts as the weld pool moves up.

Consumable-guide welding

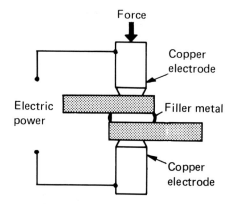

A low voltage electric current is passed through the electrodes and the workpieces. Most of the heat generated is produced in the electrodes or at the electrode/workpiece interface and is transferred by conduction to the workpieces.

Resistance brazing; resistance soldering

OTHER HEAT SOURCES

The workpiece is contained in an evacuated chamber and is bombarded by a beam of electrons produced by an electron gun operated at voltages between 5 and 100kV. The energy of the electrons is transformed into heat on their striking the workpiece. The workpiece may be out-of-vacuum (the gun only is evacuated), but within shielding for ionising radiation.

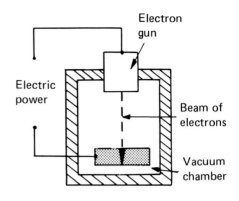

Electron-beam welding; electron-beam brazing

An alternating electric current passes through a coil surrounding, or adjacent to, the workpiece. In metals and other conductors, eddy currents are induced in the workpiece and cause its temperature to rise.

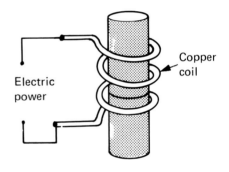

Induction welding; induction brazing; induction soldering; induction heating

The 'plastics' work to be welded is placed between metal electrodes, to which a high frequency (1-100MHz) high voltage (5-15kV) is applied. Current flows and the work is heated; the electrodes press the molten joint faces together.

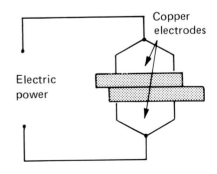

Dielectric welding; radio frequency welding (RF welding); high frequency welding (HF welding)

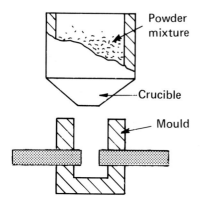

A mixture of aluminium powder and a metal oxide powder (iron, copper, etc.) is ignited in a crucible. The oxide is reduced to the metal with the evolution of intense heat, and on completion of the reaction the crucible is tapped; molten metal flows into the joint to melt the ends of the workpieces and form the weld.

Thermit welding

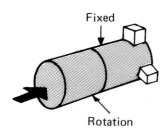

One component remains stationary while the other is rotated against it under pressure. Heat is generated at the interface by friction, and when forging temperature is attained the rotation ceases abruptly; a forging pressure is then applied to effect the weld. As an alternative to rotation, relative motion may be reciprocating or orbital.

Friction welding

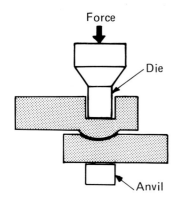

A force is applied by dies to the outer faces of the joint; the inner faces are thus brought into extremely intimate contact and a weld is effected.

Cold pressure welding

The components are heated by gas jets while under pressure, and when the correct temperature has been attained they are forged together under increased pressure to effect the weld.

Gas pressure welding

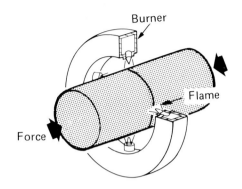

The assembled components with preplaced brazing alloy are heated in a muffle type furnace which may contain an inert or reducing gas. The furnace may be heated by gas, oil, or electricity.

Furnace brazing

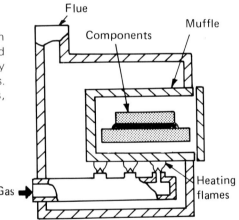

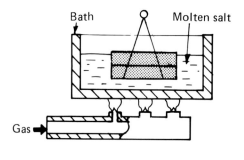

The assembled components are lowered into a bath of molten salt, flux, or filler metal, from which heat is conducted into the components. The bath may be heated by gas or electricity.

In salt bath brazing the salt acts both as a flux and as a heat transfer medium. The filler alloy is pre-placed.

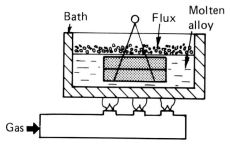

In dip brazing and soldering the molten filler alloy is covered by a layer of flux. The molten alloy provides filler metal and acts as a heat transfer medium.

Salt bath brazing; dip brazing; dip soldering

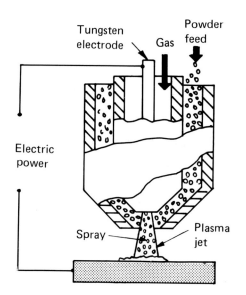

A wire or powder is fed into the nozzle of the gun where it is fused by a gas flame, an arc, or a plasma jet. The molten particles are projected in the form of a spray by means of compressed gas or air.

Thermal spraying; plastic spraying

The edges of the thermoplastic material and the filler rod are heated by a stream of hot air or nitrogen which issues from the welding torch at a temperature of 300ºC. The surfaces soften and the filler rod is pressed into the weld groove. A pressure weld is formed between the two sheets of material and the filler rod. The torch is heated by gas or electricity.

Hot gas welding

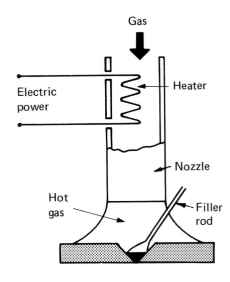

The ends of the thermoplastic components to be joined are placed in contact with a heated tool held at a temperature of about 300ºC. When the ends have become softened the tool is removed and the components are united under pressure. The tool is generally heated by electricity. To weld sheet materials heated dies are pressed against the outer surfaces and heat for welding is conducted to the joint surfaces.

Heated tool welding

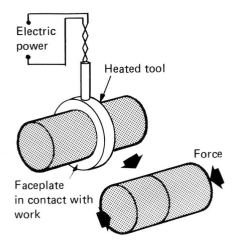

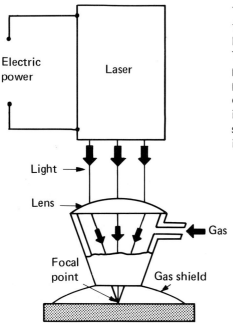

The laser generates energy in the form of a powerful parallel beam of light of one wavelength ('colour'). The lens focuses this beam to a point at which the entire output power (50W-10kW) of the laser is concentrated. When a workpiece is placed at this point it will absorb some of the light, which will raise its temperature.

Laser welding and cutting

Further information
See Refs 1.1-1.3, Appendix 1.

CHAPTER 2

GAS CYLINDERS AND GAS WELDING AND CUTTING EQUIPMENT

Flames for welding and cutting usually burn a fuel gas with oxygen for maximum combustion temperature. The fuel gas for welding is acetylene, in current practice supplied from a cylinder at a relatively high pressure. Acetylene or liquefied petroleum gas (LPG) from a cylinder is used for cutting; equipment for this process is known as high pressure plant and is the subject of this chapter. Alternatively, acetylene may be generated as required by the action of water on calcium carbide; equipment for this process is known as low pressure plant and is the subject of Chapter 3.

Chapter 4 deals with safety precautions and the chapters specifically devoted to gas welding and cutting conclude with Chapter 5 on protective clothing, eye and head protection.

HIGH PRESSURE PLANT

High pressure fuel gas from cylinders can be used (via the correct pressure regulators) for all gas welding and cutting equipment: a typical high pressure oxyacetylene welding set is shown in Fig.2.1.

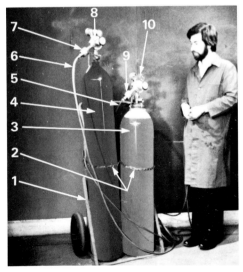

2.1 *High pressure welding set:*
 1 - *wheeled trolley*
 2 - *cylinder securing chains*
 3 - *maroon acetylene cylinder*
 4 - *black oxygen cylinder*
 5 - *red acetylene hose*
 6 - *blue oxygen hose*
 7 - *oxygen flashback arrestor*
 8 - *oxygen regulator*
 9 - *acetylene flashback arrestor*
 10 - *acetylene regulator*

17

Cylinders

Gas cylinders, Fig.2.2, are painted in distinctive colours as specified in Ref.2.1, and it is most important to ensure that only the specified gases are stored in them. To prevent the interchange of fittings between cylinders of combustible and noncombustible gases, the valve outlets are screwed left- and right-hand, respectively. Information concerning each type of gas cylinder in common use for gas welding and cutting is given below, the pressures being true at 15°C.

Oxygen: supplied in black cylinders at a pressure up to 172 bar (2500lb/in^2) at 15°C. The valve outlet has a right-hand thread and only oxygen regulators should be used.

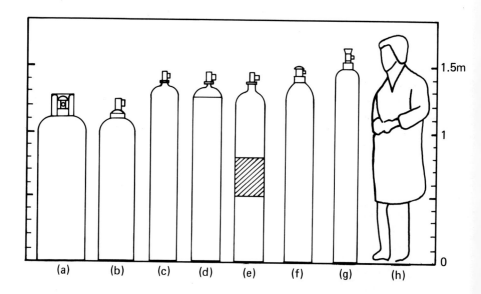

2.2 *Common welding gas cylinders: (a) propane, red, left-hand thread on outlet, (b) acetylene, maroon, left-hand thread on outlet, (c) oxygen, black, right-hand thread on this and other outlets except hydrogen; similar cylinders for compressed air (grey), argon (blue), hydrogen (red, left-hand thread on outlet), (d) nitrogen, grey with black top, (e) argon mixtures, blue with green (CO_2 mixture) or black (oxygen mixture) band, (f) helium, midbrown cylinder, right-hand thread on outlet usually to US standard, (g) CO_2, black with white stripe, right-hand male thread on outlet (other cylinders have female threads), and (h) human figure for scale. Uses of these gases cover gas welding and cutting, shielding, and other purposes*

Acetylene: supplied in maroon cylinders at a pressure of 18 bar (261lb/in^2) at 15°C. The valve outlet has a left-hand thread and only acetylene regulators should be used. Acetylene cylinders must always be stored and used in an upright position.

Hydrogen: chiefly used for heating and cutting and supplied in red cylinders at a pressure of 175 bar (2540lb/in^2) at 15°C. The valve outlet has a left-hand thread and only hydrogen regulators should be used. *Other fuel gas (LPG and acetylene) regulators must never be used* as they are designed for lower inlet pressures.

Propane: supplied in large-diameter red cylinders and stored in liquid form. The valve outlet has a left-hand thread and a propane regulator should be used with this gas. The cylinders must always be stored and used in an upright position. Propane liquefies when compressed into the cylinder, and, if the valve is not at the top, liquid may be discharged and damage the equipment.

When not in use the cylinder heads should be protected to prevent damage to the valves. Older cylinder designs had a screw-on cap, but the modern trend is to provide a shield permanently secured to the cylinder; this demands no user action, and gas from a leaky valve cannot accumulate inside, Fig.2.3. Cylinder valves and all other gas connnections should be kept free from oil or grease, dry and clean.

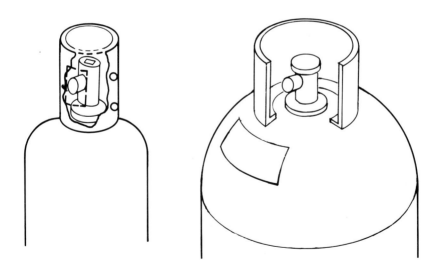

2.3 Gas cylinders with permanent valve shield: (left) hydrogen (shield partly cut away), (right) propane

Cylinders for oxygen, hydrogen, and propane are simple pressure vessels, but those for acetylene are more complex. Gaseous acetylene may explode if compressed alone; figures given for the critical pressure vary widely but UK legislation implies relative safety up to 0.6 bar (9lb/in^2), and satisfactory operation with added safety precautions up to 1.5 bar (22lb/in^2). Acetylene cylinders are therefore filled with a porous substance, such as charcoal, kapok, or kieselguhr, which is soaked with acetone; the gas dissolves safely in the acetone, Fig.2.4. Hence cylinders of acetylene are sometimes known as 'dissolved acetylene' or 'DA'. It is unsafe to withdraw acetylene from the cylinder at a rate exceeding one-fifth of its contents per hour; if this rate is exceeded acetone will be mixed with the acetylene. If a greater delivery is required, a number of cylinders should be manifolded together, Fig.2.5. A purpose-made connector and pipe should be chosen, which has been constructed of materials compatible with acetylene and pressure tested.

Gas cylinders must be treated with care and not subjected to mechanical damage, falls, or undue heating. If cylinders have to be handled by means of a crane they should be secured in a special carrier, and on no account may an ordinary chain sling be employed. A suitable type of carrier for two cylinders is shown in Fig.2.6.

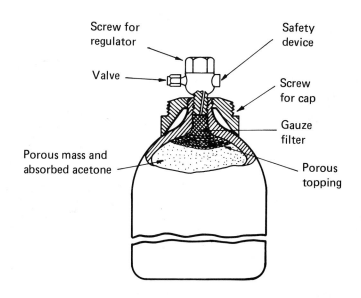

2.4 Acetylene gas cylinder

2.5 Manifolding gas cylinders

Where cylinders are moved with a fork lift truck, care must be taken to prevent them rolling off the forks by using suitable packing. The valves, too, will be exposed to risk of damage if narrow doorways have to be negotiated and special care should be taken in such circumstances.

Gas cylinders, other than those for acetylene and propane, may be stacked either horizontally or upright. In the former the stack should not be more than four cylinders deep, with the large cylinders at the bottom and safely wedged. If they are to be stored in an upright position they must be secured so that they will not fall.

Acetylene cylinders must always be stored and used in an upright position. If they are allowed to rest horizontally acetone will be withdrawn with the acetylene.

2.6 Carrier for two gas cylinders

Where portable plant is required the oxygen and fuel gas cylinders should always be transported on a suitable trolley, which should never be allowed to rest horizontally, Fig.2.1.

If a cylinder valve or outlet connection leaks and cannot be tightened with a spanner, the valve should be closed and the cylinder returned to the supplier with a suitable label attached; they should never be packed with washers.

Cylinder stores

Wherever gas welding or cutting is carried out on a regular basis it will be necessary to store supplies of gas; where usage is not on a sufficient scale to warrant a bulk tank and pipeline system, this will be in cylinders. It is desirable to have a special store set aside for gas cylinders, both because a number of safety precautions can be built-in, and because it facilitates the organised return of cylinders for refilling to maintain supplies.

Fuel gas and oxygen supplies should be kept in separate stores. The store should be of fire-resistant construction, using lightweight materials such as asbestos or metal sheet which will not act as effective projectiles in the event of an explosion. Ventilation should be provided top and bottom so that it is effective for gases which are either lighter than air (acetylene, hydrogen) or heavier (propane); to the same end there should be no sump or pit in which heavy gas can accumulate. If in the open, cylinders should be protected against direct sunlight to avoid excessive heating (in the height of summer in the UK, rather longer elsewhere.) This may be combined with roofing, again of lightweight construction, to provide protection against rain if the store is outside, or oil drips or spray if the store is within an enclosed workshop. Electric lighting should be either from fittings outside the enclosure or of flame-proof construction, to avoid ignition of any accidental escape of gas; for the same reason, no smoking should be allowed in or near the store. Cylinders should be easily placed into, and secured in, racks clearly marked to indicate gas type and whether full or empty.

Adequate access should be available for both the gas supplier's delivery transport and the user's distribution transport. The area should be used for the storage of cylinders only, and be kept clean and tidy; this can best be done by appointing a responsible storeman, who can then be trained on action in the event of emergency. A permanent notice outside the store should indicate the type and location of all the gas cylinders within, and the name and location of the storeman. Because oxyacetylene equipment is readily portable it is attractive to thieves and vandals, so adequate security measures should be taken.

If, in spite of all precautions, fire does break out, the cylinders may eventually explode on prolonged heating, so the fire brigade should be kept informed of the store location which should be clearly marked; a cooling jet of water should be played on the cylinders if possible, or, if overheating cannot be avoided, the area must be evacuated.

When planning the cylinder store it is also appropriate to consider whether a trough and water supply can be installed at a suitable distance to facilitate the emergency cooling of a heated acetylene cylinder. More positive recommendations may be found in Refs 2.2-2.5 (Appendix 1).

Bulk storage
Oxygen, propane, and several other potentially dangerous 'gases' can be bulk stored in liquid form in special storage tanks in the user's works; the tanks are refilled from the gas supplier's road tanker. Acetylene cannot be supplied in this way, and manifolded cylinders must be used.

The distribution pipelines must meet stringent standards regarding leakage, protection from damage, and fittings such as flashback arrestors. Acetylene pipelines must also be free from pure copper and silver, either in the pipe itself or in silver solder for fittings; Ref.2.2 gives details, or see under 'Fittings' below.

The safety precautions needed in the installation of both storage tanks and pipelines are usually met by entrusting the work to a specialist firm, responsible to the gas supplier, who will need to be assured of its suitability before commencing bulk supply.

Bulk storage and pipeline distribution reduce the handling required for cylinders, improving safety and reducing costs; in fact, bulk storage may be economic even at low rates of use.

Regulators
Regulators must always be fitted to the cylinders to reduce the gas pressure from cylinder pressure to the working pressure of the blowpipe, and only regulators designed for the gas being used may be fitted to the cylinders. Simple needle-valves are not permissible because they will not prevent pressure in the blowpipe and hose-lines rising each time the control valves are closed or if the nozzle becomes blocked, nor will they prevent a reverse flow of gas towards the cylinder.

The adjusting screw of the regulator must always be released before the cylinder valve is opened, which should be done slowly. The cylinder valve must be closed before the regulator is removed. The regulators should be detached from the gas cylinders unless they are being conveyed on a trolley or other special truck.

Further information on the construction, care, maintenance, and repair of regulators may be found in Refs 2.6 and 2.7.

Gauges

Reference 2.8 lists the precautions needed in respect of gauges, and should be consulted before attempting to use a gauge supplied for another purpose. The most satisfactory solution is to use only those gauges supplied for the purpose.

Blowpipes

Blowpipes of the high pressure and injector types may be used with high pressure equipment, Fig.2.7. Blowpipe tips should never be cleaned by a drill: only a hardwood stick or the reamer supplied for the job should be used for this purpose.

Blowpipes should be regularly checked to ensure that they have suffered no damage and are in good working order. If repair is needed, or in any event at regular intervals, they should be dismantled and cleaned, preferably by the maker or a firm specialising in this type of work. Further information may be found in Ref.2.9.

Hoses

The blowpipe should be connected to the regulator by means of the special canvas-reinforced rubber hose, supplied to Ref.2.10; red hose should be used

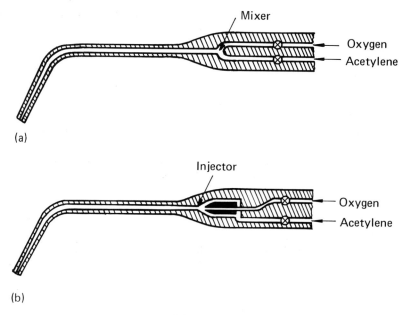

2.7 Blowpipe principles: (a) high pressure, (b) low pressure

for acetylene and hydrogen, orange hose with nitrile rubber inner lining for propane or other LPGs, and blue for oxygen.

The hose should be securely attached to its connections by means of suitable 'single-shot' hose-clips (worm drive clips or wire binding should never be used); attention to this will prevent accidents arising from the hose blowing-off from a fitting. When it is necessary to couple two lengths of hose together the special couplers available should be used. The hose should be inspected frequently for cuts or cracks and, if these are found, the defective hose should be renewed.

Fittings
If fittings have to be made up for the acetylene supply it is most important to ensure that copper or copper-rich alloys are not used. Copper in contact with acetylene is liable to form a dangerously explosive substance called copper acetylide. Only metal containing less than 70% of copper should be used.

If fittings are silver-soldered the solder should contain no more than about 40% silver and 20% copper: suitable grades are AG2 or AG3, to Ref.2.11.

Where lengths of hose are made up with end fittings supplied separately, particular care must be taken to ensure that right-hand thread fittings, with plain hexagon nuts, are always used for the blue oxygen hose, and left-hand threads, with nicks on the corners of the hexagons, for fuel gas, as specified in Ref.2.12.

On no account should any oil, grease, or other fatty substances be allowed to come into contact with oxygen regulator valves or fittings, as these substances are spontaneously combustible in the presence of oxygen and a fire or explosion may result.

Valves and fittings for all purposes should be kept scrupulously clean, and care should be taken to make certain that no grit or other foreign matter is allowed to remain on them. Attention to this small point will save much trouble arising from leaks, and will prevent the buildup of any dangerous concentrations of gas. The assembled equipment should be checked for leaks with a 0.5% detergent solution, Fig.2.8.

Hose check valves
If the blowpipe tip becomes blocked, gas may feed from the hose carrying the gas at the higher pressure back down the other hose. This can be obviated to some extent by setting both regulators to give the same outlet pressure, but, since this 'backfeeding' produces an explosive mixture in one hose, it is customary to provide extra protection. This is usually done by making the hose connectors at the blowpipe with integral nonreturn valves, known as 'hose check

valves'. If hoses are extended or repaired, valves should always be fitted at the blowpipe inlet or within one metre of the inlet connection.

2.8 Hose check valves being tested for leaks: upper nut, plain hexagon (oxygen); lower, nicked (acetylene)

Flashback arrestor
Should a backfire be propagated back into a hose from the blowpipe it is possible for it to continue back towards the regulator. If it reaches the acetylene cylinder it may internally fire it, as acetylene may either polymerise or decompose in the cylinder without external oxidant, and eventually it could explode. Although the risk is small, where it is not easy for the worker to reach the cylinder and turn off its valve should a backfire occur, a flashback arrestor should be fitted at the regulator outlet. In one typical unit, the pressure surge of a backfire triggers a spring-loaded valve to cut off the supply; to restore the supply when the cause of the backfire has been removed, the operator must operate a lever or other mechanism on the valve to reset the spring, Fig.2.9. The main application of this device is on the acetylene line, although a second in the oxygen line will give added security. Alternative designs of flashback arrestors are available and suppliers will be able to advise on the choice of a suitable unit.

Equipment purchase
All gas equipment should be purchased from suppliers who are prepared to undertake that it has been manufactured to comply with British Standards or other appropriate specifications. Some equipment imported into the UK has been found to be noncomplying and dangerous.

2.9 Flashback arrestors: (left) older-established design, reset by lever, (right) newer development, reset by disconnecting inlet and pushing pin into gas passage, preventing attempts to reset it under pressure

Further information
Gas and equipment suppliers will readily provide instruction and advice on the safe use of their products; other sources are listed in Ref.2.13, Appendix 1 (Published information) and Appendix 2 (Useful organisations).

CHAPTER 3

LOW PRESSURE PLANT

As an alternative to the supply of acetylene from cylinders, described in detail in the last chapter, it is possible to generate acetylene as it is required by the action of water on calcium carbide:

$$CaC_2 + 2H_2O \rightarrow C_2H_2 + Ca(OH)_2$$

Although this technique is almost obsolete in the UK it may be found in use where acetylene is not available in cylinders, and the information in this chapter will be of use in such circumstances.

The terms low and medium pressure are used in conjunction with acetylene generating plant and refer to the following ranges of pressure:

Low pressure: up to 0.16 bar (2.5lb/in^2)
Medium pressure: above 0.16 bar (2.5lb/in^2) but not exceeding 0.6 bar (9lb/in^2)

The generated gas is piped to the welding or cutting blowpipe, to which oxygen is fed from cylinders.

All concerned with the installation and operation of acetylene generators should be fully conversant with the maker's instructions regarding the working and maintenance of the generator and its auxiliary equipment before any attempt is made to operate the plant. It is recommended that a copy of the maker's instructions should be conspicuously posted in the generator house or on the generator itself.

Stationary generators
Diagrams of the two types of generator are shown in Fig.3.1. Stationary acetylene generating plant should be installed either in the open air or in a well-ventilated building away from the main workshops, and the generators and fittings should be maintained in good order.

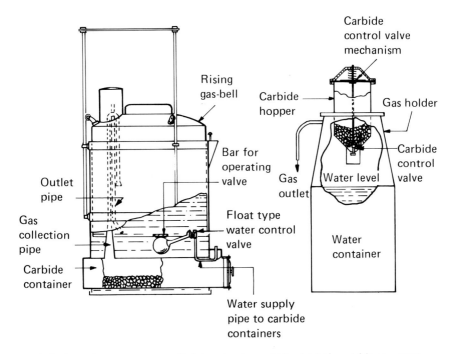

3.1 Acetylene generators: (left) water to carbide, (right) carbide to water

The ventilation of the generator house is extremely important and should be such as to prevent the risk of an explosive or toxic atmosphere being formed. Adequate lighting should be provided. Electric lamps should be located outside the building so that their light may pass through sealed glass windows. Switches and other electrical gear should not be in the house. Smoking, flames, torches, welding plant, or inflammable materials must be excluded from the house or the vicinity of an open-air generator.

When it becomes necessary to dismantle a generator for repair, all the carbide must be carefully removed and the plant should be filled with water to ensure that every part of the generator is free from gas. The water should be left in the plant for at least half an hour, after which it may be drained off and work commenced. It is preferable that this work should be carried out by the makers of the equipment or by a specialist.

When a generator is being recharged or cleaned it is most important to ensure that none of the old charge is used again: only new carbide should be used, as a serious explosion risk is present when partly spent carbide is replaced in the generator. If pieces of carbide have become wedged in the feed mechanism or have adhered to other parts of the plant, they should be removed carefully to avoid sparking; preferably nonsparking tools made of bronze or other suitable nonferrous alloy should be used.

Portable generators
Many of the precautions outlined above apply also to portable plant. This plant should be used, cleaned, or recharged only in the open air or in a well-ventilated shop and away from any inflammable material. The generators should be lifted only by means of the special lifting eyes provided.

Back-pressure valves
It is essential to make sure that a properly designed back-pressure valve, which can be regularly inspected internally, is fitted between the generator and each blowpipe to prevent a backfire, or a reverse flow of gas, from reaching the generator. The valve should be inspected after each backfire and the water level checked daily. A section of a typical valve is shown in Fig.3.2.

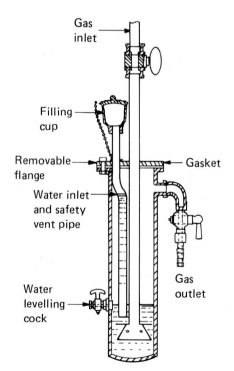

Blowpipes and fittings
Where the acetylene supply is in the low pressure range, only blowpipes of the injector type designed for low pressure operation should be used, because attempts to use the high pressure type will be accompanied by dangerous back-firing. The first backfire may displace the water from the back-pressure valve, and, if this is not noticed, a subsequent backfire may pass the valve and reach the generator. Medium pressure acetylene supplies may be suitable for some standard equal pressure blowpipes, otherwise the precautions concerning these and other fittings already dealt with in Chapter 2 apply.

3.2 Hydraulic back-pressure valve

The storage of calcium carbide
Carbide is supplied in sealed drums, in which it should always be stored before use. The chief concern in storing carbide is to keep it completely dry, and the stores must always be situated under cover to protect the drums from rain. Any dry, well-ventilated building may be used (including the generator house); if it adjoins another building it is necessary to ensure that the party wall is fireproof.

The store room should be suitably ventilated to prevent an accidental accumulation of acetylene. The ventilation of the store is of paramount importance and must be provided through the roof. The carbide should be stored on a platform raised above floor level to ensure circulation of air beneath the drums.

A prominent notice should be posted on the store:

CALCIUM CARBIDE
NO NAKED LIGHTS OR
SMOKING IN THE BUILDING

The drums of carbide should only be opened immediately before the generator is charged, and the special opener provided should always be used. A hammer and chisel must not be used because of the risk of an explosion. It is dangerous to leave carbide drums exposed to rain; moreover, drums which have been rolled through rain or even through surface pools of water are liable to explode when opened.

Water must never be used to extinguish a carbide fire. Buckets of dry sand should be kept available for the purpose. Fire extinguishers of the dry powder, halon, or carbon dioxide types are also satisfactory.

The storage and use of calcium carbide in the UK is subject to restrictions and licensing, of which details are given in Refs 3.1-3.3 in Appendix 1, and a summary in Chapter 28.

Disposal of calcium carbide residues
Note that spent calcium carbide will continue to evolve acetylene in hazardous quantities for some time after removal from the generator. For safe disposal, it should be spread in a thin layer on a secure open-air site provided with suitable warning notices, and preferably irrigated with a water spray until all reaction has ceased.

CHAPTER 4

SAFETY PRECAUTIONS DURING GAS WELDING AND CUTTING OPERATIONS

In common with the other welding and cutting processes, gas welding and cutting is quite safe if elementary precautions are taken. Elimination of danger from welding and cutting is more often than not a matter of the application of sensible precautions: carelessness can so easily lead to personal injury or damage to property. The equipment is often used by, for example, maintenance workers who have not received specific training; in such circumstances rigorous supervision and control of portable equipment is essential. Supply hoses should be arranged so that they are not likely to be tripped over, cut, or otherwise damaged by moving objects: a sudden jerk or pull on the hose is very liable to pull the blowpipe out of the operator's hands, cause a gas cylinder to fall over, or a hose connection to fail.

EXPLOSION RISKS
Explosions can occur when acetylene gas is present in air in any proportion between 2 and 82%. Acetylene is also liable to explode when under excessive pressure, even in the absence of air. The first essential requirements are, therefore, adequate and proper ventilation, and the examination of the installation to ensure that it is free from leaks.

Explosions in the plant itself may be caused by 'flashback', which results from dipping the nozzle tip into the molten pool, mud, or paint, or from any other stoppage at the nozzle; the obstruction so formed causes the oxygen to flow back into the acetylene pipe and communicate ignition back towards the generator or cylinder. Any particles of slag or metal that become attached to the tip should be removed, and if the blowpipe tip becomes hot when working in a confined space or close to a large mass of hot metal it should be cooled frequently by immersion in a bucket of water after extinguishing the flame.

The importance of careful handling of compressed-gas cylinders cannot be too strongly emphasised, and it is most dangerous to use them as supports for the work or to allow them to remain near furnaces or other sources of heat. Should an acetylene cylinder become heated accidentally, or become hot as a result of excessive or severe backfire from the use of faulty equipment, the gas manufacturers recommend that it be dealt with promptly as follows:

'Shut valve, detach regulator, remove cylinder outdoors at once, spray with water to cool, keep cool with water. Leave outdoors. Advise suppliers immediately, quoting cylinder number where known.'

Should a generator, valve, regulator, or any other piece of equipment become frozen, it should be thawed out by means of hot (but not boiling) water; no other method should be employed to thaw equipment.

Special precautions are necessary for the repair of vessels that have held combustibles; details of the recommended procedure will be found in Chapter 9.

Mixtures of propane and air in any concentration between 2 and 9% are also liable to explode. Special care is needed as propane is heavier than air and can collect in sumps, tanks, etc. If it enters a drainage system, it can follow the fall of the pipes or gullies and travel distances of the order of kilometres until it reaches a source of ignition. If this hazard is present it will be preferable to use acetylene, which is lighter than air and will rise away from the work area.

If working in an enclosed space cylinders should not be taken inside, to avoid hazards from leaks in cylinder connections or regulators. Equipment should be withdrawn during work breaks to prevent a buildup of gas from a leak, at the blowpipe for example.

A further protection against the risk of igniting an explosive mixture in the equipment itself, additional to the hose check valves mentioned in Chapter 2, is the practice of purging. Before lighting the blowpipe, fuel gas and oxygen must be allowed to flow for a few seconds (or more for long lengths of hose) separately through the systems to the blowpipe tip, ensuring that each gas line (regulator, hose, etc.) contains only its own gas, and not a mixture, regardless of the previous history of the equipment.

If leaking fuel gas catches fire, the cylinder valve must be closed to extinguish the flames. This is facilitated by observing the following rules:

1. Open the valve only half a turn
2. Leave the key in position in the acetylene cylinder valve, or chain it to the set for prompt location in emergency
3. Have an asbestos glove available for use where the burning gas is close to the valve

FIRE RISKS
The risk of any welding or cutting operation starting a fire is considerable, and so this general problem is the subject of Chapter 26. However, there is one risk special to gas welding, and particularly to gas cutting, which will be dealt with here.

Oxygen enrichment
Normal air contains only some 21% of oxygen; the remainder, mainly nitrogen, takes no part in most combustion reactions and so slows down the burning by simple dilution.

If the oxygen content is increased, burning intensity and speed is increased, normally nonflammable materials may burn, and oil or grease may catch fire spontaneously. Oxygen may be released into the air by leaks in equipment, by supplies being left on, or by excessive purging. In the normal operation of the flame cutting process about 30% of the oxygen supplied is released unconsumed to the atmosphere: gas cutting should never be undertaken in a confined space without proper ventilation arrangements (see Chapter 24). Note that, although fuel gases are treated with odorising agents, oxygen is odourless, and workers may not notice dangerous concentrations. It is very dangerous to search for gas leaks with a naked flame; only a weak (0.5%) solution of detergent in water should be used for this purpose. It is best to avoid the use of soap solution as this may react with oxygen when it dries out.

Further information is given in Ref.4.1.

EYE INJURIES AND BURNS
During welding and cutting operations precautions must be taken to prevent burns of the eyes and exposed parts of the body which may occur as the result of spattering of incandescent metal particles and from flying slag particles. The intense radiation from the flame and incandescent metal in the weld pool can cause considerable discomfort to the operator and others in the vicinity of the operation, unless the precautions described in Chapter 5 are taken.

The operator's body and clothing must be adequately protected from sparks, flying particles of incandescent metal, or slag. No oily or greasy clothing of any kind should be worn.

Articles which have been welded will be very hot on completion, and it is recommended that these should always be clearly marked HOT to warn other employees who may have to handle them. The marking should be removed when the article is cool enough to be handled without injury if the most effective protection is to be achieved. In practice, a reasonable rule would seem to be that everything on a welding bench should be treated as hot, and that articles which are not in areas protected by ropes or barriers should be marked. Small piece parts may be placed in a marked container. If it is thought that an article may be hot it may be approached with the back of the hand, which is sensitive to radiation from a very hot object; if a minor burn is sustained, it will not be as painful or disabling as one on the palm.

FUME RISKS
Good ventilation must always be provided for gas welding (see Chapter 24). The heat produced by prolonged contact of the acetylene flame with a large mass of metal will lead to the formation of oxides of nitrogen, and in confined spaces special ventilation or breathing equipment will be necessary. The fumes given off when welding and cutting parts which have been galvanised, lead-coated, or otherwise treated, may be injurious to the operator, and special precautions must be taken. The powder-cutting process used to cut stainless steel and non-ferrous metals also requires special precautions.

In the above operations the operator should guard against the possibility of inhaling toxic fumes, for example, by wearing a suitable respirator: reference should be made to Chapter 24 for further details.

LIGHTING
A good general standard of illumination at the work is essential in welding shops and booths (see Chapter 25).

Further information
See Ref.4.2 for a general review.

CHAPTER 5

PROTECTIVE CLOTHING, AND EYE AND HEAD PROTECTION, FOR GAS WELDING AND CUTTING

Gas welding and cutting operations emit much heat and light so that protection is required for the body, eyes, and head.

PROTECTIVE CLOTHING FOR THE BODY
Where molten metal or hot particles are emitted during welding the welder should wear a leather apron to protect his clothing. In cutting operations it is advisable to wear leather spats to prevent hot particles from falling into the boots or shoes. Reference 5.1 is a detailed specification for protective clothing. A well-protected gas cutter is shown in Fig.5.1.

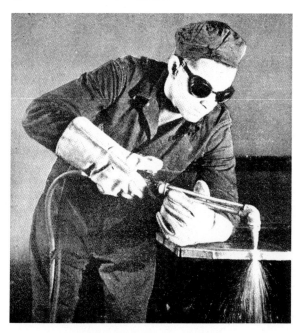

5.1 Well-protected gas cutter

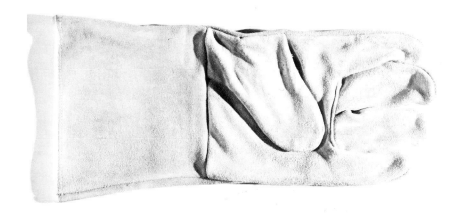

5.2 Welder's glove

The clothing, e.g. boilersuit or overall, worn under such protective clothing should be of material which will not burn easily: for example it may be of a dense weave, or be treated with fireproofing chemicals.

Gloves
For protection against heat, spatter, slag, etc. during normal gas welding operations the type of glove shown in Fig.5.2 is satisfactory. Reference 5.2 is a detailed specification.

For the repair of large castings the gauntlet type of glove alone is not sufficient and, to give protection from the intense heat, it has in the past been recommended that an asbestos cloth jacket should be worn or, alternatively, an apron with sleeve protectors of this material should be used.

However, at the time of revision of this text (1980) there is a widely held opinion that workers should not be exposed to asbestos in any form, especially where there is an increased risk of inhaling dust. At present the best course appears to be, in existing applications, to check exposure to toxic fibres and, in new applications, to try conventional leather jackets and aprons, to modify the job in some way to improve the lot of the worker, or to use mechanised or robotic welding.

5.3 Gas welder's goggles

5.4 Gas welder's goggles suitable for wear over spectacles

PROTECTION FOR THE EYES AND HEAD

In gas welding operations it is generally unnecessary to wear a helmet as in electric arc welding, and suitable goggles will be found satisfactory to protect the eyes from the heat and light radiated from the work. (For further details on the function of filter glasses see Chapter 8.) A typical pair of gas welder's goggles is shown in Fig.5.3 and in Fig.5.4 a type suitable for wear over spectacles is illustrated.

The goggles are provided with a filter sheet and a clear protective cover sheet on the outside, Fig.5.5. The cover sheet should be cleaned as required and replaced when scratched or damaged to such an extent that it is difficult to see through; the sheets may be of glass or heat-resistant plastic. The filter should be selected in accordance with Table 5.1 which is extracted from Ref.5.3. For slag removal operations a pair of goggles fitted with clear safety-glass lenses should be worn.

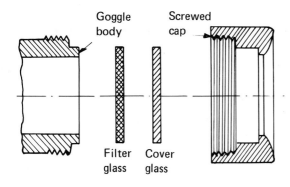

5.5 Filter and cover glasses, exploded sketch

For the repair of large castings it is good practice to use a handshield or helmet of the type described in Chapter 8 to protect the face from the intense heat radiated. A helmet of the type described in Chapter 24 which feeds a fresh air supply to the welder's face can considerably ease the physical strain on the welder.

Table 5.1 Recommended filters for gas welding

Welding process	Filters required for welding	
	Without flux	With flux
Gas welding aluminium and aluminium magnesium alloys; lead welding or oxyacetylene cutting	3/GW	3/GWF
Oxygen machine- and hand-cutting Oxygen gouging Flame descaling Silver soldering Fusion welding zinc-base die castings Bronze welding light gauge copper pipe and light gauge steel sheet	4/GW	4/GWF
Fusion welding copper and copper alloys Fusion welding nickel and nickel alloys Fusion welding steel plate All bronze welds in heavy gauge steel and cast iron, except preheated work. Rebuilding work of relatively small steel parts and areas for fusion welding. All hardfacing operations, including rail resurfacing	5/GW	5/GWF
Fusion welding heavy steel. Fusion welding heavy cast iron. Fusion welding and bronze welding preheated cast iron and steel castings. Rebuilding large steel areas, e.g. large cams etc.	6/GW and 7/GW	6/GWF and 7/GWF

CHAPTER 6

THE CARE OF ARC WELDING AND CUTTING EQUIPMENT

Arc welding and cutting equipment may be divided into two groups:

Equipment connected to an electricity supply
Engine-driven equipment

EQUIPMENT CONNECTED TO AN ELECTRICITY SUPPLY
With stationary transformers or motor generator sets it is recommended that a suitable switch-fuse be mounted adjacent to the equipment so that it may be isolated from the supply main, if necessary. Portable sets having trailing cables should be provided with interlocked fused-switch sockets and plugs at the supply end of the cable to give protection to the trailing cable as well as to the equipment. Should damage to the cable occur it can thus be isolated, as can the equipment, by means of the switch-fuse. Use of the interlocked type of socket will ensure that the equipment is not plugged in or disconnected while on load. Where lengthy cables are in use, a switch or other means of cutting off the supply close to the operator in the event of emergency may be required; if there is no switch on the set itself it may be desirable to fit one at that end of the cable (see also page 42).

The cases of both portable and stationary sets must be earthed to provide protection against live-to-case faults, although it should be understood that no connection should be made between mains earth (primary side) and secondary connections to the welding circuit. The workpiece should be earthed by an earth lead independent of the return lead, Fig.6.1. The earth connection must be capable of carrying the full secondary current without damage; cable similar to that of the welding circuit will be suitable.

The oil level in transformers and oil-immersed reactors should be checked from time to time, and the oil tested to ensure that no moisture is present. Most electricity supply authorities will be in a position to help with the testing since they will have oil testing and filtering plant at their establishments; otherwise it will be best to leave well alone if facilities are not available.

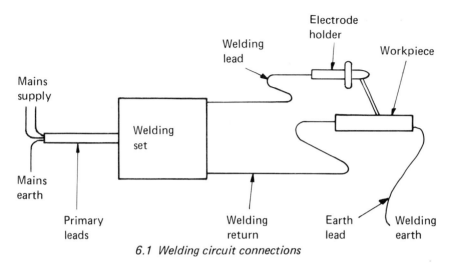

6.1 Welding circuit connections

Rotary equipment should be 'blown out' from time to time, particularly where there is metallic dust in the atmosphere of the shop in which the equipment is installed. The equipment must first be disconnected completely from the mains supply. The operation can be carried out using either a bellows or reduced-pressure dry compressed air, care being taken to ensure that the nozzle is not brought too closely into proximity with the windings of the coils and other delicate electrical components.

Similar cleaning is desirable for air-cooled transformers, and particularly for the semiconductor rectifiers used for DC supplies.

ENGINE-DRIVEN EQUIPMENT
Engine-driven equipment should be sited so that no danger to health arises from the exhaust gases. Care should be taken to ensure that the plant is on a level platform, and that the brakes have been applied or chocks fitted under the wheels to prevent movement from vibration. In very cold weather the unit should be positioned so that the prevailing wind is not blowing into the radiator. All mechanical guards should be in position, and a satisfactory means should be provided to start the engine.

Engine maintenance should be carried out in strict accordance with the details given in the maker's handbook, the generator should be blown out from time to time, and all electrical connections checked and tightened.

Particular care should be taken with regard to fuel leaks: all accumulations of fuel should be cleaned off and all connections made good. Care should be taken to avoid spilling the fuel when filling the tank.

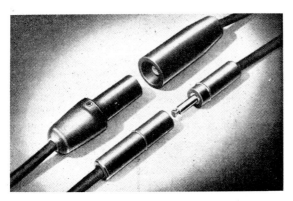

6.2 *Arc welding cable connectors (to Ref.6.2)*

ALL EQUIPMENT AND WELDING CIRCUITS

In service, the welding operator should check all external connections daily, and should report any weaknesses, defects, etc. that he may find. A periodic inspection should be carried out by a responsible person nominated for the task. He should ensure that all connections are clean and tight, that they are correctly made, that the correct types and sizes of cable, earthing clamps, Fig.6.3, electrode holders, Fig.6.3, cable connectors, Fig.6.2, etc. are being used, and he should particularly ensure that the earthing arrangements are satisfactory in all respects. Any connection which is hot to the touch after use should be dismantled, cleaned as necessary, and re-assembled.

The size and construction of cables suitable for welding circuits is shown in BS 638 (Ref.6.1). Permissible currents for each type of cable at given duty cycles are tabulated. When applying these, note that it may be necessary to use larger cables than those indicated to reduce the voltage drop to an acceptable figure, and that cables should be allowed free ventilation; for example they should not be coiled up while in use.

Particularly with manual metal-arc (MMA) or similar processes where the electrode holder is permanently live, it will be necessary to consider where the circuit can be broken should a welder receive a shock and be unable to let go. Connectors to BS 638 (Ref.6.2), Fig.6.3, can be disconnected live in an emergency, or it may be possible to switch off at the set itself or at the mains supply to it. The disconnection point should be readily accessible and reasonably close to the operator; 10m is suggested as a reasonable distance but this will depend on local conditions.

ELECTRODE HOLDERS

BS 638 (Ref.6.3) specifies three grades of electrode holder (at the time of writing, a revision is scheduled):

Grade A: fully insulated, suitable for the most adverse conditions, e.g. shipbuilding

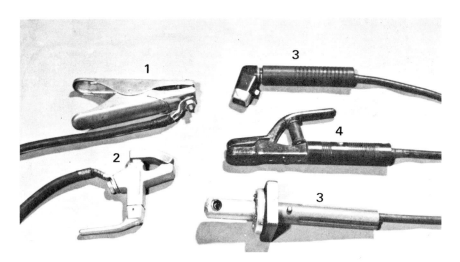

6.3 Arc welding accessories: (left) welding return and earth lead connectors: 1 — spring clamping; 2 — screw clamping; (right) electrode holders, all fully insulated: 3 — screw clamping; 4 — spring clamping

Grade B: fully insulated, suitable for general shop application but particularly for welding in enclosed conditions, e.g. inside vessels
Grade C: semi-insulated for general open shop conditions

Grade C is rarely used nowadays, as modern insulating materials allow holders meeting grades A and B to be produced without difficulty.

The electrodes are held in the holder by screw or spring pressure. Screw fixing is more effective and ensures a good contact but spring clamping allows electrodes to be changed much more quickly, Fig.6.2.

Discomfort to the operator can be reduced by effective heat insulation on that part of the holder which is held in the hand, and to obtain this some types employ air spaces or ducts along the handle.

It is important to ensure that the cable connection is a good electrical joint and so designed that the continued flexing of the cable will not cause wear and failure of the insulation.

The jaws and connections of electrode holders should be cleaned and tightened periodically to prevent overheating.

ACCESSORIES
Operators' handshields and helmets should be checked from time to time to ensure that there is no damage, such as burn-holes, which would result in light being admitted, or would provide a conducting path to the inside. A check should also be made that an unbroken filter of a suitable shade is fitted (see Chapter 8), and that the cover sheet is present and reasonably free from spatter.

Protective clothing, such as aprons, gloves, spats, etc., should be inspected for burst seams or holes through which molten metal or slag might enter. Welding booths and portable screens should also be checked regularly to ensure that there is no damage which might result in the arc affecting persons working nearby.

GAS-SHIELDED ARC WELDING
The equipment used for inert-gas arc welding is more delicate than that employed for conventional arc welding, and rather more care should therefore be taken in handling the plant. A greater number of cables and tubes are necessary and these may be kept tidy by taping or tying them together.

METAL-ARC GAS-SHIELDED WELDING
It should be noted that the consumable electrode wire is 'live' to the workpiece and earth; therefore an insulated reel or holder must be provided, and accidental contact of the wire with earthed metalwork, such as the frame of the set, avoided whenever the arc power supply is likely to be switched on.

TUNGSTEN-ARC GAS-SHIELDED WELDING
To strike the arc, a high frequency (HF) unit is used to provide a spark from electrode to the workpiece. The workpiece terminal on the HF unit should be connected by the return lead firmly to either the bench on which small jobs are placed, or to the job itself if this is standing on the floor. An earth lead to the bench or workpiece is essential. Earth, return, and torch leads should be kept reasonably short to avoid untoward effects such as stray sparking and to obtain satisfactory arc initiation.

Rubber hoses and rubber-covered cables must not be used anywhere near the HF discharge, because the ozone which it produces will rapidly rot the rubber. Equipment makers can supply hoses and cables made from suitable plastics: some plastics, such as PVC, conduct high frequencies to a certain extent, and so rapidly break down. Rubber-covered cables are generally satisfactory for the primary connections, and may be acceptable for work return and earth leads if these do not have to run near the arc.

The carbon content of some black rubber hoses for gas or water supplies can cause leakage conduction of HF currents. These and other effects can cause sparking and other unwanted behaviour, with no obvious conducting path. The HF supply cable to the torch must have special insulation to avoid stray sparking. The HF on its own will not cause an electric shock, but may startle the unwary when a spark jumps on to the hand; it often produces deep burns which are slow to heal.

Dirt and metallic or other conducting dust can quickly cause a breakdown in the HF discharge unit, which should be blown out regularly to prevent such dust accumulating.

CHAPTER 7

SAFETY PRECAUTIONS DURING ARC WELDING AND CUTTING OPERATIONS

Arc welding is a safe operation when carried out under normal and correct workshop conditions, but it must be pointed out that equipment free from defects and a well-arranged, well-lighted, properly ventilated, tidy workplace are important factors in safe working, Fig.7.1.

7.1 Welding shop curtained off into bays. Services (such as electricity and CO_2 supplies, and extractor hoses) run along fixed partitions. Curtains can be slid back on rails when larger work area is required. Individual benches have pillar and clamp for safe workholding

46

The types of accident which occur in normal operation, through carelessness or faulty equipment, are set out below.

ELECTRIC SHOCK

Under normal industrial conditions a supply of no more than 110V has been used to obtain an adequate measure of safety against electric shock, but this is centre-tapped so that the effective voltage to earth is only 55V; a reduction to 49V as the maximum voltage considered safe to touch is proposed in the European Economic Community harmonisation programme.

In welding, the voltage to earth may be higher. The open-circuit voltage (OCV) of MMA welding equipment, for example, is usually 80V AC, one side earthed. Also the welder is liable to come into contact with live conductors in the welding circuit in the course of his work. As the risk of shock from 80V is normally low, welders may come to regard this as safe and become careless. When work is carried out in hot or damp surroundings, the risk of a fatal shock may be significantly increased.

Direct current welding supplies are usually slightly lower in voltage, around 70V. For auxiliary supplies (such as wire-feed motors), many makers are adopting 42V to meet the European regulations.

No-load, low voltage devices

For work carried out in confined spaces or dangerous positions growing use is now being made of electrical devices which can be installed in the welding circuit so that only a low voltage (25V) is present at the holder when welding is not taking place. The devices usually contain contactors which bring the main welding current into circuit as soon as the electrode is placed on the work. Once the arc is struck the voltage in the circuit automatically falls to arc voltage, which is usually of the order of 25-30V and is therefore safe.

The devices are also known as low voltage safety devices, or OCV reduction relays. It may not be possible to use these in conjunction with remote-control current-adjusting systems.

When installing such a system, effective maintenance must guarantee continued correct operation. If the device fails to open, a welder who has become used to handling the electrode holder with impunity will receive a severe shock; if it fails safe, there will be a temptation to cut it out of circuit and resume work.

Safe working practice

The risk of shock should never be ignored. However, in certain situations, e.g. in cramped spaces such as boilers and small tanks which may be warm and damp, or in insecure positions where a shock may lead to a serious fall, extra care (such as that enforced under a permit-to-work system) is necessary; in particular, the following points require attention.

The work must be securely earthed, separately from the welding return connection. With multiple-operator MMA installations extra care is needed to ensure the integrity of the connections, to avoid the increased risk of shock from the higher voltage (up to 160V) between phases. Advice on a suitable earthing point may be obtained from the electricity supply authority. All insulation must be in good condition on electrode holders, guns, torches, cables, and accessories.

The operator should see that his working position is dry, secure, and free from dangerous obstructions, and that he is using adequate protective clothing, as described in Chapter 8.

If damp or sweaty conditions are unavoidable, an extra check should be made to see that equipment with the lowest available OCV is used: some DC generators have an OCV that is adjustable down to about 50V. Ideally, a no-load low voltage device as described above should be used. The welder should be kept under surveillance when he is handling live parts, particularly if the working area is cramped and inaccessible. A means to cut off the current must be available near the operator for use in emergency.

There should be adequate protection (guard rails, safety harness) if there is any risk of a fall.

The casing of mains-operated equipment must be securely earthed to the mains earth.

Ideally, the electrical power supply should be switched off when not in use, but this is not always practicable in MMA welding so some provision should be made to accommodate the electrode holder when not in use. The practice of laying a live holder on a face screen or on a pair of gloves is not sufficient, nor is it satisfactory to suspend it by the welding cable in such a manner that it may come into contact with other equipment. An insulated hook should be provided on which it may be safely placed, or a fully insulated holder should be used.

Gas-shielded arc welding process
Care should be taken to avoid dragging the cables and plastic tubes across hot plates or welds. Where HF currents are used, cables should not be allowed to coil as this would reduce the spark. The torch lead should not come in contact with the job or any other 'earth', as this will put an extra strain on the insulation.

Particularly in the metal-arc gas-shielded process, accidental depression of the torch or gun switch can lead to unwanted arcing, especially where the set remains switched on when the trigger is released. Safe stowage will avoid this hazard, and also those of fire or burns from a hot gun, or of damage to the torch.

Electric heating
Where work is suspended by slings or chains care is needed to avoid welding current flowing in them, overheating and weakening them; overhead cranes may also have their motors burnt out by excessive welding currents flowing in their windings, either directly or by induction. It is best to ban welding on suspended work, but if it must be carried out a separate welding return taken direct from work to set return connection, and a sound earth lead attached directly to the work, will minimise the problems.

Portable electric tools
Where earthed portable electric tools are in use near welding equipment, welding without a secure welding return connection to the work may fuse the earth lead of the tool, removing the protection it gives against electric shock from a breakdown in the insulation of the tool. It is better practice to use double-insulated tools without an earth lead, thus avoiding the problem completely. Portable tools used in the welding shop should be checked frequently, with particular attention to earth lead or double-insulation as appropriate. Earth leakage circuit breakers in addition will provide an extra measure of safety.

First aid
The first aid treatment for electric shock causing cessation of breathing is artificial respiration, which is extremely effective if applied immediately (see Chapter 27). To avoid delay it is desirable that all those working in areas where there is risk of electric shock should be trained in the application of artificial respiration.

Further information
A more detailed treatment of electrical safety in arc and other welding processes is given in Ref.7.1. Heart pacemakers — see Appendix 4.

BURNS
Burns are almost invariably caused through lack of care, or through failure to wear protective clothing. The operator should make sure that his clothing is free from oil or grease, and that his workplace is tidy and not encumbered by any inflammable material which may be ignited by sparks or spatter. He should also avoid working with his sleeves rolled up and his hands and forearms unprotected by suitable gloves or sleeves. In overhead welding, a protective cape is a useful addition to his equipment.

It must be remembered that the skin can be burnt by the rays emitted from the arc as well as by contact with hot metal. It is therefore important to see that no parts of the body which might otherwise be exposed to these rays are unprotected.

Articles which have been welded will be very hot on completion, and it is recommended that these should always be clearly marked HOT in chalk or other suitable material to warn other employees who may have to handle them.

Gas-shielded welding processes
In both tungsten and metal-arc gas-shielded processes more ultraviolet light is emitted than in MMA welding at the same current, so extra protection for the skin may be needed, for example for the back of the neck.

Tungsten-arc gas-shielded welding
A HF spark is used to initiate the arc and to stabilise the AC arc during welding. The spark can cause small, deep burns if concentrated at a point on the skin, e.g. through a hole in the insulation of the torch. It is also important to feed in filler wire along the material being welded, as this earths the wire and eliminates any risk of electric shock.

EYE INJURIES
If the eyes are exposed to the light of the arc, even for quite short periods, arc eye may develop; this can be avoided by using a headshield or helmet fitted with a suitable filter (Chapter 8) and by avoiding stray flashes from other welding arcs. Adequate screening to protect workers in the vicinity is essential. If any person is exposed to a flash, the effects of arc eye may be minimised by the immediate use of special eye lotion which should be available in the first aid box or ambulance room of the factory; a sufferer should report for treatment as soon as possible (see Chapter 27).

Eye injuries can also occur during deslagging operations; Chapter 8 should be consulted for recommendations on protection against these.

If eye strain is to be avoided an adequate standard of lighting is essential; this will also contribute to the making of a sound weld, because the operator can see that all traces of slag have been removed between runs (see Chapter 25).

Painting arc welding booths
The use of black paints for the inside of welding booths has become an acceptable practice, but, provided a matt finish is achieved, almost any colour is equally effective. Pastel shades are less depressing to look at, and may reduce eye strain by improving the illumination at the work. Emulsion paint is cheap, easily applied, and readily available in matt finishes and suitable shades.

Curtaining arc welding booths

Curtains can be used to close off the entrance to arc welding booths where other workers would be exposed to radiation from the arc. Heavy fireproof material, usually a light green in colour, is available for the purpose. It is fire-resistant to a certain extent, but an effort should be made to keep arcs, flames, and hot metal away from it as far as possible.

Recently, a transparent tinted material has been on sale which allows the work to be viewed from outside. As it is claimed to meet similar standards to normal filters this should be safe for the viewer, but surface reflections may distract the welder inside.

Screens

All electric welding operations should be screened to prevent the rays of the arc from affecting other persons working in the vicinity. Where the work is carried out at fixed benches or in welding shops permanent screens should be erected; where the nature of the work is such that these are not practicable, temporary screens should be used to limit the radiation.

All screens should be opaque, of sturdy construction to withstand rough usage, and of material which will not readily be set alight by sparks or hot metal. They should not, however, be so heavy or cumbersome as to discourage their use. A suitable construction is hardboard carried on a light tubular steel framework, emulsion painted to a suitable colour (see above).

Information on the distances from the arc at which ultraviolet radiation is reduced to the US daily TLV is reproduced in Table 7.1 from data in Ref.7.2.

FUME RISKS

In addition to fumes evolved from electrodes and fluxes, coated or otherwise treated metals may give off toxic fumes. Special care in ventilation must also be taken in welding nonferrous metals and certain alloy steels. The ultraviolet light from the arc, particularly the gas-shielded arc, may form ozone from the air or phosgene from solvents. Shielding gases are released into the atmosphere, and may build up where ventilation is restricted, or even in open-topped tanks, as they are generally heavier than air. Fume problems are discussed more fully in Chapter 23, and their solution by such means as fume extraction in Chapter 24.

FIRE RISKS

Sparks and spatter from the arc are always liable to ignite any inflammable material in the vicinity. Care should therefore be taken to make sure that the workplace and the surrounding area are clear of anything which may catch fire; the precautions required are discussed in Chapter 26.

The precautions which must be taken when welding vessels which have contained combustibles are set out in Chapter 9.

Table 7.1 Distance from arc at which ultraviolet radiation is reduced to US daily Threshold Limit Value for various exposure times, from data in Ref.7.2

Process	Parent metal	Shielding gas	Current, A	Distance (m) for		
				1min	10min	8hr
Manual metal-arc	Mild steel	-	100-200	3	10	70
Metal-arc gas-shielded	Mild steel	CO_2	90	0.9	3	20
			200	2.1	7	50
			350	4	13	90
(flux-cored wire)	Mild steel	CO_2	175	1.2	3.5	24
			350	2.2	7	50
	Mild steel	95% argon + 5% oxygen	150	3	9	65
			350	6.5	20	140
	Aluminium	Argon	150	3	10	70
			300	5	17	110
	Aluminium	Helium	150	1.3	5	35
			300	3	10	70
Tungsten-arc gas-shielded	Mild steel	Argon	50	0.3	1	7
			150	0.9	3	20
			300	1.6	5	40
	Mild steel	Helium	250	3	10	70
	Aluminium	Argon	50 AC	0.3	1	7
			150 AC	0.8	2.7	18
			250 AC	1.3	4	30
	Aluminium	Helium	150 AC	0.7	3	20
Plasma arc welding	Mild steel	Argon	200-260	1.6	5	33
		85% argon + 15% hydrogen	100-275	1.7	5.5	40
		Helium	100	2.9	9	65
Plasma arc cutting (dry)	Mild steel	65% argon + 35% hydrogen	400	1.3	4	30
			1000	2.5	8	55
Plasma arc cutting with water injection	Mild steel	Nitrogen	300	3.2	11	75
			750	1.8	5.5	40

CHAPTER 8

PROTECTIVE CLOTHING, AND EYE AND HEAD PROTECTION, FOR ARC WELDING

The welding arc gives off light and heat of high intensity, so that the following types of protection are essential for the welder:

Protective clothing for the body
Protection for the eyes and head

In addition, workers engaged in the vicinity of welding operations should be protected by means of screens erected around the place of welding. Welders' assistants should wear suitable goggles and gloves where necessary.

The information given below will also apply in general to arc cutting operations.

PROTECTIVE CLOTHING FOR THE BODY
It may be necessary for the welder to wear a leather apron or other suitable devices to protect his body and clothing from the heat of the work, to prevent burns which can be caused by small globules of metal falling on his thighs and legs, and, in gas-shielded arc welding processes, to avoid the destructive action of ultraviolet radiation on clothing. The need for an apron will depend upon the nature of the welding process and the position in which the operator is welding relative to the work. An apron is particularly important where the operator is seated for welding at a bench. If he is wearing an overall of flame-proofed cloth, without turnups at cuffs or trousers, further protection is often unnecessary. However, if the welder is wearing ordinary clothes and welding in certain positions, leather or asbestos sleeves and leather spats, as well as his apron, may be necessary. A suitable cap is important when welding in the overhead position.

Standard items of protective clothing for welders are specified in Ref.8.1.

A fully protected arc welder is shown in Fig.8.1; note, however, that the sleeve protectors and spats should be worn only when necessary as the welder may be exceedingly uncomfortable in the hot environment of welding, and, in high ambient temperatures, may collapse from heat stroke.

8.1 *Fully protected arc welder*

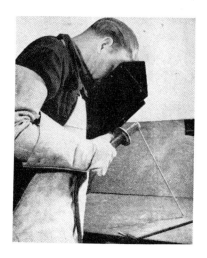

8.2 *Welder's protective sleeve*

Gloves

To protect the hands against heat, spatter, and radiation, gloves must be worn, and should be of the leather gauntlet types with canvas or leather cuffs, as specified in Ref.8.2. Alternatively, gloves with shorter cuffs together with separate sleeves should be worn. An inseam type of glove with reinforcement between the thumb and forefinger is preferable. The seams should be inside, so as not to trap globules of molten metal which would burn the stitches. Rubber or plastic gloves are not suitable.

A typical welder's glove is shown in Fig.5.2 and the use of a protective sleeve can be seen in Fig.8.2.

PROTECTION FOR THE EYES AND HEAD

Head protection

For all arc welding work, either a helmet, Fig.8.3, or handshield, Fig.8.4, is essential to protect the welder's head from radiation, spatter, and hot slag. The handshield protects one hand as well as the face; it is the least tiring protection to use but gives the least protection to the head. For the gas-shielded arc welding process, flat handshields provide insufficient protection from reflected radiation. The helmet is supported on the head to protect the face, top of the head, and

8.3 Welder's helmet   *8.4 Welder's handshield*

the throat of the welder. For work on aluminium structures, or when working very close to other welders, protection for the back of the head should be arranged, otherwise light from another welding arc reflected from sheet metal may burn the back of the neck or the scalp.

Reference 8.3 specifies helmets and headshields, and includes a requirement that they shall be electrically insulating to provide shock protection when working in confined spaces.

All helmets and handshields are provided with a filter and a protective cover sheet on both the inside and out; the cover should be cleaned as required, and replaced when scratched or damaged to such an extent that it is difficult to see through. The filter should be selected as recommended later in this Section.

When removing slag after welding, clear goggles, a clear face shield, or clear safety spectacles should be used to protect the eyes from injury: slag particles are often hard and sharp, and will be hot if removed immediately after welding.

Eye protection
The arc used in welding and, to a lesser extent, the molten metal in the weld pool radiate light and heat. Filters are used to prevent this radiation (a term which covers ultraviolet, visible light, infrared, and heat emissions) from damaging the eyes. The filter glass is so-called because it separates or filters these different kinds of radiation and allows only a limited amount of one kind to pass

through; ideally, only visible light should do so, but it is not possible to tell by looking through a filter glass which types of invisible radiation are being stopped.

If too much ultraviolet radiation is transmitted by a filter the welder will get a sunburnt face and arc eye; the latter is an intense irritation of the eye occurring some hours after exposure. If too much infrared radiation is passed the welder will feel his face become uncomfortably hot and, in course of time, after repeated exposure, damage to the eyes may occur. If too much visible light is passed the welder will be dazzled and will not see the arc comfortably, but if too little light is passed he will not see his work well enough; he will therefore suffer from eye strain, which may lead to headaches. Reference 8.4 discusses these matters in detail.

BS 679 (Ref.8.5) specifies the transmission of light through filters suitable for use in welding and cutting. A number indicates the amount of visible light passed: lower numbers correspond to a lighter filter. For each shade number, ultraviolet and infrared must be cut down by a specified factor more than the visible.

A wide choice of density is available for each class of work, and if the welder can comfortably see what he is doing and is using a filter which complies with BS 679, his eyes are safe. The recommendations of this Standard for arc welding are summarised in Table 8.1 for easy reference, but the Standard itself should be consulted for details. The 'auxiliary heat-absorbing filter' referred to there is not widely available, but there have been some attempts to introduce a metallised filter which reflects most of the incident infrared radiation.

Where two or more shade numbers are recommended for a particular process and current range, the higher shade numbers should be used for welding in dark surroundings and the lower shade numbers for welding in bright daylight out of doors.

The higher emission of infrared radiation in gas-shielded welding processes gives rise to an uncomfortable increase in the temperature of the filter, and it is recommended therefore that an auxiliary heat-absorbing filter be placed between the cover glass and the filter glass.

The intensity of visible radiation in the tungsten-arc process is less than in the metal-arc process for comparable currents, but it is recommended that the same degree of protection against infrared radiation is used.

Where the surface temperature of the filter may rise above $100^\circ C$, e.g. when welding in a preheater furnace, it is advisable for filters of solid glass, or glass laminates enclosing dyed interlayers, to be used.

Table 8.1 Recommended filters for electric welding

Welding process	Approximate range of welding current, A	Filter(s) required
Manual metal-arc	Up to 100	8/EW
		9/EW
Automatic metal-arc	100-300	10/EW
		11/EW
	Over 300	12/EW
		13/EW
		14/EW
Metal-arc gas-shielded	Up to 200	10/EW
		11/EW
Carbon-arc	Over 200	12/EW
		13/EW
Atomic hydrogen		14/EW
Metal-arc gas-shielded	Over 500	15/EW
		16/EW
Tungsten-arc gas-shielded	Up to 15	8/EW
	15-75	9/EW
	75-100	10/EW
	100-200	11/EW
	200-250	12/EW
	250-300	13/EW
		14/EW

Electronically switched filters: a type of filter has been recently introduced which is switched from clear to filtering condition by a photocell which, responding to ultraviolet radiation on the front of the mask, controls an electronic circuit which energises a liquid crystal element. The advantage claimed is that a clear view of the work is available during nonarcing periods.

Contact lenses: a story about the effect of an arc fusing a welder's contact lenses in place has been reprinted a number of times, causing anxiety; however, no corroborative detail has been forthcoming. The UK Employment Medical Advisory Service of the Health and Safety Executive has issued an authoritative medical opinion on the subject which states that a contact lens will probably reduce the amount of ultraviolet light entering the eye, and therefore reduce the risk of arc eye. If a contact lens wearer feels the symptoms of arc eye coming on, he should remove the lenses and not insert them again until he is better; if he does this, there should be no further ill effects.

Associated workers: where it is not practicable to screen welding operations completely it may be necessary to provide eye protection for those working nearby. According to circumstances this can be achieved with a light shade (3) of filter to BS 679, or with a filter which simply cuts out the ultraviolet radiation, leaving the visible unaffected, and therefore appears clear. The filters do not require the elaborate mounting and protection of those to be used by the welder himself, and simple one-piece moulded goggles are available, either green to BS 679, or clear; most clear plastic materials absorb ultraviolet light, so normal clear protective goggles may be satisfactory, but this should be confirmed with the supplier before a final choice is made. Ordinary glass spectacles also significantly absorb ultraviolet, protecting their wearers to some extent against arc eye. Patterns of clear safety spectacles are available which can be worn over normal spectacles, protecting them and the wearer from spatter and mechanical damage.

Reference 8.6 contains data on the safe working distances from arcs to ensure meeting US Standards for the exposure of the unprotected eye, and an extract is given as Table 7.1 of the previous chapter.

ARC CUTTING
The precautions to be taken during arc cutting operations do not differ materially from those for arc welding. However, since the currents used are generally higher, and there is a larger amount of molten metal to cope with, more and heavier protective clothing is generally necessary. When the carbon-arc cutting method is used, the arc is larger than the normal metal arc, and the use of a good helmet is recommended in preference to a handshield. Spats of leather or asbestos are also advisable to protect the feet from the falling molten metal.

With air-arc cutting in particular there is an extremely high noise level, and the operator should use suitable ear protection.

EAR PROTECTION IN ARC CUTTING AND OTHER PROCESSES
Air-arc cutting or gouging produces high noise levels, and therefore ear protection is required to prevent those exposed to the noise (mainly the operators) from temporary or permanent partial deafness.

Ear muffs

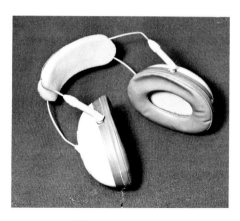

8.5 Ear muffs

Ear muffs are widely used in industry generally, but some problems may be encountered in applying them to welding. For welding, the makers do not recommend liquid-filled seals, which ensure a better fit and hence better sound sealing, as spatter could perforate the seal and release the liquid. Models are marketed with a rear band which fits round the back of the neck clear of a helmet headband, but these may be found uncomfortable and it may be possible to select a normal headband model which will fit under a helmet. Another critical point is whether the cups are shallow enough to clear the helmet. A good point in favour of muffs is that they can be seen to be in use, Fig.8.5.

Ear plugs

Ear plugs are also effective, offering similar attenuation and overcoming the problems of compatibility with the welder's helmet. Instruction to the wearers should emphasise that the ears and plugs should be clean before insertion to avoid the risk of ear infection.

Plugs can either be re-usable rubber or plastic mouldings, or disposable glass down, which ensures that a clean plug is used each time and also allows for visitors to be issued with hygienic ear protection.

Other processes

Oxygas and oxyarc cutting and gouging, particularly in heavy gouging or on thick plate, can produce high noise levels. But if a mechanised cutter is in use the operator may stand far enough away not to be exposed to too great a total noise energy. A safe rule is to check noise level if a problem is suspected or any temporary deafness is encountered.

Metal-arc gas-shielded welding with dip transfer is comparatively noisy, though rather below the present suggested legal limit of 90dB(A) unless welding in an awkward position; the helmet may provide some protection.

Further information

Reference 8.7 gives Tables from which the likely effects of exposure to a given noise level can be expected estimated on a statistical basis, that is, the percentage of persons with a given hearing loss. Reference 8.8 gives general guidelines.

CHAPTER 9

PRECAUTIONS FOR WELDING AND CUTTING VESSELS WHICH HAVE HELD COMBUSTIBLES

It is dangerous to weld a vessel which has *at any time* contained a flammable liquid unless proper precautions are taken, and work should not be attempted in the absence of proper facilities to ensure safety. An explosive ignition of vapour may be caused by the arc or flame used in welding, cutting, and brazing, or even by a hot soldering iron. The danger is present not only in vessels that have held volatile liquids such as petrol, but also in those which have contained liquids such as tractor vaporising oil, diesel fuel, paraffin, linseed oil, concentrated aqueous ammonia, etc., Fig.9.1.

9.1 Gas cutting this tank containing residue of fuel oil caused an explosion which killed three men (Ref.9.1) (Copyright The Times, London)

If a vessel has to be worked upon and its previous contents are unknown, it should be treated as if it had contained a flammable substance, however long it may have remained empty.

The methods of preventing accidents of this kind are of two main types:

Removing the flammable material
Making the material nonexplosive and nonflammable

The precautions which must be taken will be discussed under two headings, i.e. those relating to small vessels, and to the large vessels that the worker will enter. It is the task of management to provide a safe place of work, with safe access, and a safe system of work, for example by appointing and training a responsible person to supervise the preparation of vessels for welding and their testing, or by organising a permit-to-work system. It is the duty of welders and other workers to carry out the orders given by management to ensure their safety: for example, to follow a permit-to-work system they should question any unclear instruction bearing on safety, and report any incidents which may indicate an unexpected hazard.

SMALL VESSELS

Making the material nonexplosive and nonflammable

In this method the air in the vessel is replaced by an inert gas or by water during the whole time in which the work is going on; the 'inert gas' may be steam, nitrogen, or carbon dioxide. Adequate displacement of air must be checked and certified by a competent person with a proprietary explosion test meter before work starts, and, for lengthy jobs, at intervals during the operation. A responsible person should be appointed by the management and receive the necessary training to assure himself that an agreed and proven procedure has been carried out, and that tests have been properly carried out and certified.

Openings in the vessel will have to be plugged in such a manner that air cannot readily flow in, and to prevent any substantial excess pressure building up.

A proprietary method of filling a tank with inert gas is to use the gas to blow a foam which is fed into the tank. In the UK this inert-gas foaming is available as a service including testing and certifying the atmosphere, and renders practical a number of operations where an inert gas on its own would escape or disperse. It is a tried and tested procedure and is suggested as a first choice. The service operator takes over responsibility for preparation, testing, and certification.

The method of replacing the air in the vessel with water is sometimes applicable when it is possible to weld below the waterline, or it is safe to leave a small air space immediately below the part to be heated. The air space will still contain flammable vapour, and this method should be used only where the air space is free to vent to the outside atmosphere. A typical application of this method is the repair of the joint between a filling pipe and a fuel tank, Fig.9.2. The following methods should *never* be used:

Washing out the vessel with cold or hot water, or allowing water to run through it
Blowing out the vessel with compressed air
Cleaning by means of trichloroethylene or carbon tetrachloride

(However, blowing out with low pressure compressed air may be used to remove the explosive hydrogen-oxygen mixture formed in lead-acid and other secondary batteries, especially during charging, before welding the lead of the terminals; this must be done carefully to avoid any risk of acid spray. Note that repairs should not be carried out in charging rooms, in which naked lights are prohibited.)

Removing the flammable material
If the method of first choice — inert-gas foaming (see above) — is not available, the second choice is to remove the flammable material either by steaming out the vessel or by boiling.

For steaming, the filler cap and drainage plugs should be removed from the vessel; any tools used for this purpose must be of a nonsparking type, such as those made of bronze. The vessel should then be emptied and placed in such a position that condensed steam may readily drain away. Low pressure steam should be admitted, with sufficient outlets to prevent any pressure building up. A typical setup is shown in Fig.9.3.

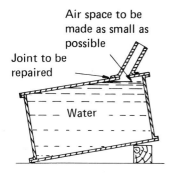

9.2 *Welding over water*

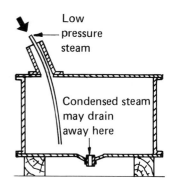

9.3 *Steaming a vessel*

Steaming should be continued until the atmosphere in the vessel is shown to be nonexplosive using an explosion test meter. Where there is previous experience relative to the type and size of job concerned, the steaming time used previously can be regarded as a minimum for future operations.

For boiling, the vessel should be fully opened up (using nonsparking tools), emptied as far as possible, and then immersed in boiling water. It is sometimes an advantage to use water containing an alkaline degreasing agent, but this agent should be of a kind that will not corrode the vessel (it will, of course, be necessary to ensure that no toxic fumes will be produced during subsequent welding). The boiling should be continued for at least half an hour, and for longer periods when necessary.

After steaming or boiling the vessel should be carefully examined and tested by a responsible person and, if acceptable, marked as ready for repair.

Further information on work on small vessels is given in Ref.9.2.

LARGE VESSELS
A welder may have to work inside a large vessel in which there is a risk of the atmosphere being explosive or toxic, or both. No welder should be required to enter any such vessel until it has been certified by a responsible person as safe for entry, and also safe to work in with an arc or flame.

Consideration must be given not only to the actual contents of the vessel but to any modification which is likely to arise as a result of the work carried out, such as vaporisation of residues producing toxic fumes.

Where a toxic atmosphere cannot be cleared from a vessel it is possible to use breathing apparatus to protect the workers who have to enter it. See Chapter 24 for further discussion, particularly regarding the case where the atmosphere is sufficiently toxic to put life at risk if equipment should fail.

The precautions which have to be taken to make such vessels safe are similar in principle to those described for small vessels, but they have to be carried out on a much larger scale. The preparation of the vessels will usually be outside the scope of the welder's work. The operator should remember that, for reasons explained earlier, oxygen should never be used to 'sweeten' the air in the vessel, and blowpipes should not be left inside when they are not in use.

PERMIT TO WORK
If a vessel forms part of an installation in use, it will often be covered by a permit-to-work system, as in oil refineries, chemical plants, and power stations.

This requires an authorised signatory to issue a permit to work before any repairs, modification, or maintenance are carried out on any part of the plant. The welder's duty is then limited to understanding and observing strictly the conditions laid down, obeying the agreed handover procedure on change of shift or completion of work, and reporting promptly any unexpected incidents which might indicate an unsafe situation, such as pipes found to be under pressure when disconnected.

Further information is given in Refs 9.3 - 9.6.

CHAPTER 10

PLASMA ARC PROCESSES

PLASMA ARC
The 'transferred' or 'direct' arc system is used, under the name of plasma arc, for most cutting, welding, and gouging. The principal use of the 'nontransferred' arc is for spraying metals and other materials, including reinforced plastics, with a variety of metallic or refractory substances such as alumina, zirconia, tungsten carbide, etc. In this instance the arc is contained wholly within the head. Metal or ceramic spraying is achieved by the introduction of wire or powder into the torch, so that it passes through the arc and is converted into a plastic or fluid state. The plastic or fluid material is then carried out of the torch by the gas stream and impinges on the component to be coated with such force that a firm bond results. The nontransferred arc is also used for gouging. A major use with either a transferred or nontransferred arc is in welding thin materials, usually referred to as microplasma welding.

Hazards
Fundamentally, the plasma process is a development of the arc principle as applied in electric welding and cutting, and presents broadly similar health hazards, with some additions.

PHYSICAL HAZARDS

Electrical
Cutting equipment requires OCV's from 100 to 400V. Although equipment should be safe under normal operating conditions, the operator must realise the danger which is present in the use of such voltages. In addition, the technique of firing the arc by a HF pulse involves the danger of exposing the operator to the dangers of an unpleasant shock and a painful penetrating HF burn.

Normally, the supplies will be safely confined within the torch. The operator should be instructed to take particular care to maintain correct welding return and earth connections to the work, to check the torch for possible damage before switching on, and never to attempt to clean the orifice by inserting a wire.

Radiation
Filters from the /EW series of Ref.10.1 provide adequate eye protection. With cutting, full face protection is required. For microplasma work, at currents up to 15A, a normal helmet is not necessary and a lighter type of shield, covering the face only, will be adequate, with a light filter, shade numbers 3-9 usually being chosen. Some types can accommodate magnifying lenses in addition to the filters, which may reduce eye strain in workers whose sight is less than perfect. A darker filter (higher shade number) will be required for operation in the same melting mode as normal tungsten-arc gas-shielded welding, at 15-100A. At currents of 100-400A the keyhole mode of full penetration welding is obtained. Both here and for cutting, the plasma arc emits considerably more radiation than the normal open arc, requiring darker filters and extra care to avoid accidental exposure. Some relevant data will be found in Table 7.1. For cutting, see also under 'Water sprays and immersion' below.

Heat
The nontransferred arc is contained within the torch, which consequently emits a hot gas jet even when it is not applied to the work. Operators must be instructed to put down the torch in a safe place or, better still, switch off before laying it down.

Noise
The high speed at which the plasma jet leaves the torch can cause it to emit intense HF noise, especially in cutting and spraying applications, and the operator must use some form of ear protection (see Chapter 8).

TOXIC HAZARDS

Gases
Adequate exhaust ventilation around the work should be provided since dangerous concentrations of oxides of nitrogen may be formed, especially if nitrogen is used in the shielding gas. The necessity for exhaust ventilation is further emphasised by the fact that ozone may also be generated.

Dust and fumes
Thermal and ceramic spraying may well present health hazards from the inhalation of dangerous dusts and fumes, depending on the nature of the material being sprayed. Consideration must also be given to the possibility of fires and explosions from accumulations of dust in the workshop. Therefore, the methods of dust and fume removal outlined in Chapter 18 'Thermal spraying' should be instituted for spraying by the plasma arc process whenever they are considered necessary. In the event of doubt, seek expert advice.

Water sprays and immersion
Modern plasma cutting installations are often equipped with a water spray covering the work area, or with a shallow tank of water in which the work is submerged. The water is effective in reducing noise, radiation, and fume.

PROTECTIVE EQUIPMENT
Besides adequate face protection, gloves must be worn to protect the operator from sprayed particles or from molten metal during cutting. In general, exposed clothing should be nonflammable.

CHAPTER 11

ELECTRO–SLAG WELDING AND CONSUMABLE–GUIDE WELDING

TOXIC HAZARDS

The fluxes in common use may be of the fused, bonded, or mechanically mixed type, the last being mixtures of several fused or bonded fluxes. The fused fluxes are the products of fused oxides and halide salts; the bonded fluxes consist of a mixture of finely divided oxides of manganese, aluminium, silicon, zirconium, and titanium, bonded together with a suitable binder and agglomerated.

Much less fume is evolved than with most other processes, since only very small amounts of flux are actually consumed. Consequently, it is safe for the process to be operated in a large well-ventilated workshop without additional precautions, but exhaust ventilation should be provided in a small inadequately ventilated location.

PHYSICAL HAZARDS

If the copper shoes are in correct position there is little risk of injury to the operator because no splash or spatter occurs. As a precaution against accidental spillage of flux or molten metal, caused by incorrect fitting of the shoes or by maladjustment of the welding controls, operators should wear protective aprons, gaiters, and appropriate footwear.

As the process is 'arc-less', except for the first few minutes when the arc is largely submerged under the flux and is also shielded by the copper shoes, there is no visual hazard for personnel working in the vicinity. The operator, however, is required to check periodically the depth of the slag bath and should be protected from glare and any possible splashing of the slag or molten metal by suitable eye or face protection. In addition, during cooling some slags splinter and fragments fly off with such force that they may injure the eyes and skin of the operator. An appropriate mask and filter may be selected from the normal welding range, or of the pattern used in the steelmaking industry, with filters to Ref.11.1.

CHAPTER 12

RESISTANCE WELDING

In general, resistance welding is carried out by the application of a powerful squeezing force to the work being welded at the same time as a very heavy electrical current is passing through the material to heat it to the necessary temperature.

The most common hazards likely to be encountered in the operation of these processes are:

Electrical
Burns
Lacerations and crush injuries
Flying particles

ELECTRICAL HAZARDS
There should be no danger from shock during the normal operation of a machine, because the voltages that are applied to the exposed and touchable current-carrying electrodes seldom exceed 20V. One side of this supply is normally earthed through the machine frame. These machines usually operate from a 400V supply which is dangerous if handled. As a precaution against touching terminals at this voltage, no covers should be removed or doors opened for inspection or maintenance purposes until the electrical power has been switched off at the mains. Doors are sometimes provided to give access to tap changing connections and control cabinets. Such doors should ideally be fitted with safety interlocks, but these are not often found in practice. In addition to the live wiring and terminals expected in electrical control gear, the thyristors or ignitrons used to control input power and their heat sinks or cooling jackets will be live whenever the mains supply is connected, even if the control unit is set to switch off the welding current.

The high currents used give rise to correspondingly high magnetic fields, particularly in large machines of the DC or 'three-phase' (low frequency) type. It has been suggested that these magnetic fields could affect the operation of electronic heart pacemakers, used to treat certain cardiac conditions. It is desirable to warn employees and visitors of this risk by notice, and by individual advice where appropriate. (See also Appendix 4.)

There is one process — HF resistance welding — in which there are high voltages on exposed welding electrodes. The frequency of the electric current is so high, usually about 400kHz, that there is little danger of electric shock if live terminals are touched. However, small but very deep and painful burns are likely to occur where the current enters and leaves the body, and for this reason adequate precautions should be taken to prevent contact with these terminals while they are alive.

PHYSICAL HAZARDS

Burns
Sometimes no visible heat is produced during the welding operation, consequently there is a tendency to think the welded parts are not hot. Burns will result from the careless handling of hot assemblies.

Lacerations and crush injuries
Resistance welding is extensively used on guillotined, blanked, or pressed components which may have very sharp edges or frays. The incidence of cuts and lacerations can be minimised by de-fraying the components before they are taken to the welding position, and by wearing protective gloves or gauntlets.

Resistance welding machines invariably have at least one electrode which moves with considerable force; in this respect they closely resemble a power press. If a machine is operated while a finger or hand is between the electrodes or platens, severe crushing will result. The risk of crushing should be considered for each machine and a suitable means must be devised to safeguard the operator.

A recent development, applicable to a wide range of machines and with little disturbance to normal operation, uses a modified air pressure system. The initial approach is with a low pressure, which cannot produce sufficient force to cause significant injury to a trapped hand or finger. When the electrode gap has been reduced to 6-8mm, in the absence of any obstruction, the full air pressure is applied to produce the normal weld squeeze force. An alternative is a probe near the electrode which switches off the machine if it encounters an obstruction.

Particularly for larger installations, a fence or grill guard, which must be closed to complete an interlock circuit, a proximity or photoelectric guard, or pressure-sensitive mats on the surrounding floor, can provide protection on the same principles as those established for power presses.

Seam-welding machines require similar protection against nipping hazards between the wheel electrodes.

Automatic sequenced machines should be fitted with one or more emergency stop buttons within reach of the operator and anyone who may be trapped.

Flying particles
Particles of hot or molten metal will not fly out of spot, seam, or projection welds if the material and welding conditions are ideal, but such particles are frequently expelled in production work, and the greatest danger is to the eyes of the operator or a passer-by. Loose metal parts should not be left in the throat of the machine because they are liable to be projected from it with some velocity by electromagnetic forces.

The flash welding process inherently produces a considerable quantity of flying red hot particles which may travel up to 6m and which may enter unprotected eyes and exposed skin with some force. Nonflammable screens should be used to protect persons in the vicinity and precautions should be taken against an outbreak of fire.

Whenever necessary, the operator must be provided with goggles, gloves, and protective clothing.

TOXIC HAZARDS
If the articles to be resistance welded are free from dirt, oil, and other extraneous material, little fume should arise during the operation. There are exceptions, however, for instance components covered with a thin layer of oil or those plated with zinc or cadmium. The latter is extremely toxic and is often used on parts for light mechanisms of all kinds; see Chapter 23 for further details. Paint will usually prevent the welding current flowing, and so will be cleaned off the work in the weld region.

The amount of fume is unlikely to be great and will disperse quickly in a well-ventilated shop, but in small and inadequately ventilated workplaces it may be necessary to provide local exhaust ventilation. In any event it is advisable to ascertain that the nature and concentration of any fume is not harmful (see Chapter 22).

NOISE
Certain resistance welding machines, for instance some that are used for flash welding, produce excessive noise which may be harmful to hearing. Where this arises from the machine operation rather than as an intrinsic feature of the process, for example noisy compressed air exhausts from actuator cylinders, or loose transformer laminations, it will generally be preferable to rectify these at source; otherwise ear protection will be required.

CHAPTER 13

THE THERMIT PROCESS

Thermit as used for welding applications is an exothermic mixture composed of finely divided aluminium and specially prepared iron oxide. Alloying elements such as manganese, carbon, molybdenum, nickel, vanadium, chromium, and titanium are added as required to give the necessary strength and hardness properties.

The essential conditions for the process are a mould surrounding the parts to be welded, a crucible lined with a suitable refractory material supported above the mould, and a metal slag container which may be sand-lined. For most welds the end of the pieces to be welded will be preheated to a temperature of up to $1000^{o}C$. Preheating torches, which are inserted into the heating gate of the mould, are fired by propane or butane and air, or by propane and oxygen.

Since the ordinary thermit charge has a high ignition point, it is necessary to place a low ignition point powder or igniter on top of the charge. The ignition powder in general use is composed of fine aluminium powder with a peroxide, chlorate, or chromate. At the completion of the thermit reaction the temperature of the resultant steel contained in the crucible is of the order of $2500^{o}C$. However, the temperature drops rapidly because of heat losses into and through the crucible lining. After the chemical reaction is complete, time is allowed for the dense steel and lighter slag to separate, and the crucible is tapped 20 to 60sec after ignition.

SAFETY PRECAUTIONS

Presence of moisture
The thermit mixture must be kept dry at all times. If it has become damp in storage, and even if it has subsequently dried out, the aluminium-oxygen balance of the mixture will have been upset; this will not only affect the quality of the thermit steel, and consequently the strength of the weld, but can cause the violent evolution of hydrogen or carbon dioxide gas. If the mixture is wet when the reaction starts, steam will be generated inside the crucible; either steam or gas will cause molten slag or steel to be ejected.

Precautions must be taken to ensure that the crucible lining is dry before the thermit mixture is added.

Welding site

The process can be operated in the open air or under cover. Nonrepetitive work should preferably be carried out under cover to protect the equipment and the job from the weather. If an existing building is used it should be of sufficient height to allow the equipment to be set up well clear of any timber structures, and if the completed job requires the use of lifting tackle, space must be available for this to be operated. Space must also be available around the job to allow free and unobstructed movement at all times for the men engaged there.

The mould

For welding of a repetitive nature where the pieces to be welded are of small dimensions a preformed refractory mould may be used. For nonrepetitive work it is usual for a steel case to be built round the job, which is rammed with a suitable grade of sand to form the mould. It is essential for the mould box to be of sufficient strength to withstand both preheating and the added weight of thermit steel at the time the weld is made; otherwise any movement or buckling of the mould box may result in a cracked mould which will allow the molten steel at a temperature of about $1800^{\circ}C$ to escape. The moulding sand must be moistened to render it workable before mould making. The mould will dry out during the preheating operation if the moulding sand has the correct permeability factor, and no trouble resulting from steam evolution should be encountered.

The 'lost wax' process is generally used in repair welding. The wax melting out produces fumes which are more of a nuisance than a hazard; but these are quickly dispersed when good ventilation exists.

The crucible support

It is usual for the crucible support to be built in the form of a bridge spanning the mould box. Tubular steel scaffolding is often used for this purpose for reasons of economy and for ease of transport, storage, etc. The structure must be of sufficient strength to withstand the weight of the crucible and thermit charge. It will be subjected to vibration caused by turbulence in the crucible during the thermit reaction, and to heat arising from waste gas issuing from vents in the mould during the preheating operation. These factors must be taken into consideration during construction. The structure can be shielded to some extent by the use of suitable deflector plates above the waste gas vents in the mould. All joints in the structure, particularly if made of tubular scaffolding, must be examined at intervals during the preheating operation and tightened if necessary.

Fire and explosion hazards
Care must be taken in choosing a welding site to ensure that fire risks, such as structures made from flammable materials, are sufficiently far from the equipment to be unaffected by splashes, or radiant heat from the crucible or mould. Buildings having wooden floors or wood in the roof structure should be avoided. The building chosen should be dry and well ventilated.

Thermit should be stored in dry conditions and be separated from the ignition powders or igniters used for starting the thermit reaction. Igniters should, if possible, be stored in a separate building. The ignition temperature of the igniter or ignition powder is about 200°-300^oC and of the thermit mixture about 1000^oC. Thermit for repair welding is normally supplied in linen bags, containing 10kg of mixture, which are packed in steel drums. The steel drums are lined with polythene and thermit should be stored in them until it is required for use. For rail welding preweighed portions are supplied in polythene bags. Thermit should always be stored away from materials of flammable nature which may be liable to self-ignition.

Some thermit mixtures and aluminium powders have been tested in the past and have been found to be a Class 1 dust explosion hazard. Because of this feature a high standard of cleanliness in the workrooms may be necessary and the accumulations of powder kept to a minimum.

VENTILATION
The equipment used for preheating may produce a toxic hazard, especially if the equipment is incorrectly used and combustion of the fuel is incomplete. In this connection the following possibilities must be avoided, i.e. leaky gas connections and the emission of waste gas or paraffin fumes. Generally, an adequately ventilated workshop should be sufficient to avoid discomfort or hazard to the operators, but, if the ventilation is inadequate and fumes are not dispersed quickly, local exhaust ventilation would be advantageous during the preheating and reaction stages.

Where thermit welding is used to join austenitic manganese steel with manganese levels of 12 to 14%, the powder contains manganese to provide a matching weld metal, and substantial toxic manganese fume is evolved during the reaction. Fume protection is essential where this type of work is carried out in a confined space such as a tunnel: it may take the form of a dust respirator of proved effectiveness (see Chapter 24).

PROTECTIVE EQUIPMENT
In rail welding the process equipment is relatively low and it is necessary for the welder to look into the crucible before tapping. He must, therefore, wear

suitable eye protection. The main radiation hazard is the infrared em molten steel. Adequate protection is given by a filter to BS 679 (Ref. 4/GW to 6/GW. A young man with good visual acuity can use a (5/GW or 6/GW) with which he gets an improved view; older men, younger men with a visual deficiency, prefer a lighter filter (4/GW): a general-purpose compromise is a 5/GW. The eye protection must also be proof against splashes of molten metal or slag and be shatterproof. These requirements may be met by wearing extra clear goggles or face shield of a pattern appropriate for molten metal service over the welding goggles. In repair welding the equipment is generally larger and assembled at greater distances from the floor. The molten metal pouring from the crucible is at about the welder's eye level, and similar filters in a helmet or handshield are recommended. Goggles are inconvenient as the welder must be able to remove his protective eye equipment rapidly to enable him to move around speedily during the operation of the process.

Protective clothing such as aprons, hoods, leg shields, and asbestos gauntlets should be provided and used as appropriate for the job.

The welder must use a clear face shield when he is trimming the weld as a protection against flying hot particles and sand.

CHAPTER 14

ELECTRON–BEAM WELDING

The essential parts of an electron-beam (EB) welding machine are a dynamically evacuated work chamber in which the work to be welded is enclosed, an electron gun, a means of traversing either the workpiece or the electron gun to follow the contour of weld, and a power pack which supplies the low and high voltages to generate, accelerate, and focus the electron beam.

ELECTRICAL HAZARDS
The safety of the equipment on the exterior of the unit is largely the responsibility of the designer and the manufacturer. To protect against electric shock the user should satisfy himself that a system of interlocks has been fitted to the various cabinets, and he must ensure that the interlocks cannot be jammed in an unsafe position. A good and reliable system of earthing is essential, linking together the working chamber, power pack, pumping units, and control cabinets.

Simple types of electron gun employ the workpiece as the anode. More refined guns have an anode constructed as part of the electron gun, the workpiece being at the same potential as the anode. For safety in operation the anode of the system is normally at earth potential. Connections made in this way ensure that the workpiece is at earth potential and that the only live parts of the high tension (HT) supply are the cathode, any subsidiary anodes in a stepped potential below earth, and the connecting cables inside the vacuum chamber.

The whole of the HT circuit should be in earthed metal enclosures and HT connecting cables should be of a suitable type, such as pliable armoured cable with proper terminals (armour clamps, glands, etc.). Such a design is adopted in most cases to ensure a reliable HT supply free from the risk of insulation breakdown; it also ensures maximum safety for the operator when handling anything external to the welding chamber. The important precaution which remains to be taken is to ensure that the equipment is totally switched off and isolated before the operator enters the chamber for the purposes of cleaning, adjusting the electron gun, or loading and unloading the assembly to be welded. A grounding rod should be used to discharge any static on the gun before working on it and to prevent any buildup during work. At all times when the chamber is at atmospheric pressure the HT supply should be isolated from the welding chamber. An effective means of securing this isolation is to ensure that the vacuum chamber is fitted with a capsule-type vacuum switch wired into a contactor which controls the supply at HT to the electron gun. In this way the HT supply is automatically isolated when air is let into the chamber.

With high volume production equipment it is not feasible to switch off the high voltage supply: instead it is 'ramped' up and down over a period of seconds; also a vacuum valve isolates the gun column from the chamber. As an additional safeguard a door switch should be fitted to the vacuum chamber wired so as to disconnect the HT supply.

The method of interlocking the doors with the electrical circuits must receive very careful consideration along the lines set out in Refs 14.1 and 14.2. Interlock switches are now available which can be overcome, by pulling out the actuating plunger, when necessary during maintenance work, and are then self-resetting when the door is closed again. Most modern equipment has the HT supply enclosed within an oil tank inside the cubicle, so there is no risk from this, but normal mains voltage hazards may still exist.

In any equipment big enough for a person to enter there is always the danger of somebody else shutting the door and switching on (or of powerful workhandling gear or door motors being energised). Protection is given if he carries with him an interlock key without which the equipment cannot be switched on (such as in the Castel key system). This can be applied both to cabinets containing electrical gear and/or to the vacuum chamber as appropriate. There is also justification for fitting a safety fence in some circumstances.

RADIATION HAZARDS

An unseen hazard associated with the use of electron beams at high voltages is the generation of X-rays. It is imperative for sufficient shielding to be provided to protect the operator from the possibility of X-rays penetrating the walls and windows of the vacuum chamber to a harmful degree. The X-radiation is more penetrating at shorter wavelengths. The wavelength is inversely proportional to the accelerating HT voltage, and the intensity of the radiation is proportional to the square of the voltage and to the beam current. It is, however, difficult to give any precise value of the voltage above which the X-rays produced are capable of penetrating the chamber. This depends not only on the voltage but on the design of the particular machine, the thickness of its chamber walls, and the material being welded. In general, for electron beams of less than 50kV, vacuum chambers which have wall thicknesses determined by structural reasons also have sufficient self-shielding. As the voltage is increased the walls of the vessel must be thickened, an economical way being the addition of a lead lining.

Any parts of the machine providing shielding against X-radiation should be interlocked in such a way that, unless they are in position, the machine cannot be energised. Should such an interlock not be practicable care must be taken that shields removed for occasional maintenance are replaced before normal operations are resumed.

Electron-beam welding machines should be checked at the time of installation for the escape of unwanted X-radiation through the chamber walls, an important possible weakness being the observation windows. It may be expected that 30kV machines will not present any hazard, the dose rate, if measurable, being less than $50\mu R/hr$ above background. Machines operating above 100kV are more likely to require precautionary measures, for example additional external shielding. The regulations of Ref.14.3 will apply to machines used in the UK. If the radiation dose rate outside the machine does not exceed 0.75mrad in air per hour, those employed using the machine will not be classified workers and will not have to wear film badges and receive periodic medical examinations. If the external dose rate does exceed this figure, an approved scheme of work will be required. However, where it is reasonably practicable the external dose rate should be reduced to less than 0.75mrad/hr. The manufacturers of the equipment will be able to advise on its likely performance in this respect, and on suitable improvements.

In the UK the Factory Inspectorate must be notified of the existence of an EB welding machine. Advice on precautions may be sought from the Factory Inspectorate, or from the Radiological Protection Service, Clifton Avenue, Belmont, Surrey.

TOXIC HAZARDS

During evacuation of the chamber the vacuum pump must remove a volume of air equivalent to the volume of the chamber. As the mechanical vacuum pump which performs this duty operates under oil for sealing and lubrication, the oil is agitated and the escaping air carries a fine mist of oil. This is unpleasant to breathe in, deposits an oil film on surfaces near the exhaust, and carries a possible toxic hazard from the major oily constituents and any minor additives. Control of this oil mist is generally desirable, and may be achieved by filtration and/or exhausting the air outside buildings, clear of any windows. Filters will need to be selected which will pass the large volumes of air involved without creating excessive back pressure on the pump. Special-purpose oil mist separators are commercially available.

The vacuum pump exhaust will carry any toxic material present in the chamber atmosphere, so it must be suitably sited and/or filtered to avoid hazard.

During welding of a metal evaporation will take place to an amount dependent upon the vapour pressure of its constituent elements. Normally the vapour will condense to form films on the interior of the vacuum chamber, but evaporation may also be accompanied, particularly if there is considerable gas evolution, by atomised particles, sometimes in the form of fine dusts.

If the materials being welded are of a toxic nature, e.g. beryllium, plutonium, etc., great care must be taken when opening the chamber to protect the operator from any dust cloud which may have arisen during the admission of air. To secure this protection it is recommended that an exhaust system be fitted to the chamber via a vacuum valve; thus the exhaust system is automatically opened to an extractor immediately the chamber door is opened. At the same time clean air is drawn into the chamber and dusts are removed in a direction away from the operator. The exhaust ventilation suggested as a possibility for the removal of fine metal dust from the chamber may well form a dust explosion hazard for certain dusts. This may lead to considerations on the siting of the associated dust collectors (if any) and the need for explosion relief. Those precautions which are usually recommended in handling metals of this nature should be taken when the assemblies are loaded or unloaded from their welding jigs. Measurements will show whether extra precautions, such as full air supply and/or dust hoods for the operators, are required for metals as toxic as those above.

14.1 Anyone trapped inside this 1m diameter electron-beam welding chamber can cut off the pumps by pulling a safety lanyard (Ref.14.4)

PHYSICAL HAZARDS

It should be borne in mind that the cooling rate of a material in vacuum is slower than if it were in air, where it loses heat not only by radiation but also by convection and conduction. Thus, if the structure being welded is not able to cool by direct conduction to its supporting jig, adequate cooling time should be allowed before handling.

The size of work which can be welded is limited only by the vacuum chamber; many chambers are large enough for a man to enter and consequently the problem of ensuring that an operator or maintenance engineer is clear of the vessel prior to evacuation must be seriously considered. There is also the need to give protection against the possible movement of work-handling gear or of a power-driven door. Therefore, on equipments of such size, locking mechanisms should be employed which make it impossible to switch on the evacuation equipment until the person who entered the chamber has come out. A warning system should enable anyone trapped in the chamber to push a button operating an emergency bell cutting off the pumps and admitting air. Very large chambers should have a lanyard-operated switch, activated on pulling or breaking the lanyard, Fig.14.1. Evacuation of this size of chamber will be slow enough to allow time for action before asphyxiation supervenes.

CHAPTER 15

LASER WELDING AND CUTTING

Recent development has established a range of welding and cutting operations which can be carried out with laser radiation, but it is not yet clear in which applications the conventional heat sources will be displaced.

It will be convenient for description of the hazards encountered to consider four classes of laser which may be used in welding or cutting, identified by their active elements:

Ruby: pulsed output (pulse duration of the order of milliseconds or less)
output wavelength in the visible (694nm, red)
active element — ruby crystal

Neodymium: pulsed output (pulse duration of the order of milliseconds or less), or continuous (YAG only)
output wavelength in the near infrared ($1.06\mu m$)
active element — neodymium in yttrium aluminium garnet (YAG) or glass host material

Ruby or neodymium lasers may be used to weld, drill or cut, or trim thick film resistors on substrates

Carbon dioxide: continuous output
output wavelength in the mid infrared ($10.6\mu m$)
power of the order of kilowatts
active element — carbon dioxide in a mixture of gases with helium and neon

Carbon dioxide lasers can be used to cut wood, plastics, or metal several millimetres thick, and to weld metal several millimetres thick

Helium-neon: continuous output
output wavelength in the visible (633nm, red)
power of the order of milliwatts
active element — helium-neon gas mixture

Helium-neon lasers are of no use directly for welding and cutting but may be used to align the optical components of the other types of laser, or large structures.

ELECTRICAL HAZARDS

The pulsed ruby and neodymium lasers draw their electrical input energy from a capacitor bank or pulse-forming network charged to several kilovolts; interlocks are required to prevent live parts being touched until the mains supply has been cut off and the stored energy safely dissipated.

Carbon dioxide lasers have a substantial high voltage supply to deliver several kilowatts at, typically, tens of kilovolts; unbeatable interlocks with back-up shorting systems are essential for safe operation.

Helium-neon lasers have a steady high voltage supply which is normally enclosed safely within the substantial case needed to protect the delicate optical components. If a flexible cable carries the supply between units it must be suitably protected and regularly inspected for damage.

RADIATION HAZARDS

The visible light from ruby lasers can enter the eye; if it does so in its usual form of a parallel beam as generated, or a nearly parallel beam reflected from a curved surface at some distance from the eye, the energy will be focused to a spot on the retina, burning it at that point. The visible light from helium-neon lasers may also cause this injury, as will the near infrared from neodymium types.

The much more intense radiation from the carbon dioxide laser is absorbed by the cornea and its moisture film, and in consequence will tend to damage the front of the eye rather than the retina. Excessive exposure to infrared is believed to increase the incidence of cataracts. The increased intensity, typically 1kW in a beam of 25mm diameter, means that skin may be burnt by the unfocused beam.

Selective filters which absorb at the appropriate wavelength have been proposed for work with visible light from ruby lasers, but they are difficult to make sufficiently effective, are expensive, and are liable to damage if they do absorb much radiation; hence safety precautions have usually been based on complete enclosure of the laser output region in a box opaque at the laser wavelength, with mechanical shutters interlocked on viewing windows. This approach is also valid for neodymium lasers in the near infrared.

As only special materials will transmit the 10.6μm radiation of the carbon dioxide laser, there is little difficulty in shielding the installation with plastic that is transparent to visible light. The complete equipment and workpiece is enclosed in a transparent plastic box with the opening panels interlocked to the laser power supply; for additional safety, operators wear clear plastic visors.

UK law (Ref.15.1) requires employers to provide suitable eye protection for employees.

Regular eye examinations may also be desirable to detect any incipient eye trouble whether it arises from a leak of laser radiation or natural causes.

Further information
Anyone working or planning to work in the field should consult the appropriate British Standard (Ref.15.2). A comprehensive treatment covering such topics as the effects of excess radiation on the eye, threshold levels, eye protection, and precautionary procedures will be found in Ref.15.3. References 15.4 and 15.5 cover US practice, and Ref.15.6 provides an extensive review of current knowledge.

CHAPTER 16

BRAZING AND BRAZE WELDING

BRAZING PROCESSES
The process of brazing (including hard soldering and silver soldering) involves joining two closely fitting surfaces by a filler metal which has a lower melting point than that of the parent metal. Health and safety hazards arise from the careless handling of the heating equipment and from the fumes which may be given off from the molten filler metal or flux.

Five basic heating methods are used in brazing:

Torch
Induction
Resistance
Furnace
Salt and flux bath

These are discussed in general terms, and then specific hazards are dealt with in detail.

Torch brazing
In this method the filler metal is either preplaced in the joint as an insert or externally applied from a rod, and a flux is used to shield the brazing operation from the atmosphere. A heating flame is produced by a torch supplied with an oxygen-fuel gas or air-fuel gas mixture, and the filler metal is fused by heat conducted from the hot component parts. Gas mixtures often used are air-natural gas, oxygen-natural gas, oxypropane, and oxyacetylene, the oxyacetylene flame having the highest temperature. In torch brazing the operators tend to bend over the job to have a closer view of the work, and hence they are liable to inhale any fumes emitted from the operation.

Mechanised torch brazing employs a number of fixed torches and the work is moved through or into the heating flames. In this method the operator assembles the components away from the brazing operation and is therefore less exposed to the fumes which may be produced.

Induction brazing
In this method, which is normally operated as a continuous process, heat is produced by eddy currents induced in the components from a HF current passing through a water-cooled coil encircling the joint; heating is intense and localised. A paste or powder flux is commonly used, the filler metal is preplaced in the joint, and the components are assembled for brazing in a jig. The operator loads one jig while another assembly is being brazed, and he is unlikely to be in close contact with any fumes produced.

Resistance brazing
Brazing by resistance heating is employed for the rapid assembly of small components; there are two methods. The first makes use of a resistance welding or heating machine equipped with special copper alloy electrodes. Heat for brazing is obtained by the resistance offered to the passage of a heavy electrical current through the joint. Heating is extremely rapid and, as the joint is enclosed and under pressure, very little fume is evolved. If the equipment is incorrectly set sparks may be emitted during passage of the current.

The other method employs two carbon electrodes which make contact with either side of the joint area, and passage of an electric current causes the electrodes to glow, thus heating the joint by conduction. Flux absorbed by the carbon electrodes may cause some fuming, but in most cases the operator does not have to bend over the work.

Resistance heating uses low secondary voltages so that there is little risk of electric shock (see also Chapter 12).

Furnace brazing
In this method the components are heated to brazing temperature in a furnace, and, although a conventional flux is sometimes used, a controlled atmosphere is normally employed. This atmosphere will often contain asphyxiating constituents and there is a risk of explosion in high temperature furnaces. Filler metals for furnace brazing do not generally contain volatile constituents so that the risk of exposure to metal fumes is small.

Salt and flux bath brazing, dip brazing
In these processes the components are lowered into a bath of molten salt, flux, or filler metal, and greater health hazards are present because the bath is kept in a molten state. A variety of salts are used, some of which are highly toxic, and appropriate exhaust ventilation should be provided. There is also a risk of spatter, especially when components are put into the bath: they must be dry to avoid a serious hazard.

EXPLOSION HAZARDS

Mixtures of some furnace atmospheres and air are explosive and it is necessary to purge the furnace of air before it is heated. There is also a risk of explosion if the door is opened when the furnace is in operation, in the absence of the protective flame. This flame is often initiated by a pilot light which must always be kept alight.

Nitrate salt baths are in common use for aluminium brazing; combinations of these salts can be explosive and care is essential, especially with dry salt. All users of nitrate salt baths should make themselves familiar with the official publications (Refs 16.1 and 16.2).

TOXIC HAZARDS

The principal hazard in brazing is exposure to toxic fumes and gases, which are generated from the constituents of the filler metals or from the coatings on the metal surface. Some general notes on these fumes follow, with sections on specific fume problems which should be taken into account when work is planned.

Provided a correct technique is followed the filler metal should be melted by heat conducted from the assembly and not directly by the flame or heat source. In this way the filler metal runs without overheating and solidifies rapidly, thus giving rise to the minimum of metal fume. The main cause of excessive metal fume in torch brazing is a poor heating technique in which the operator directs the flame at the filler metal before the parent metal is at brazing temperature. Extra heat must then be applied to bring the components to brazing temperature so that the filler may run into the joint. This procedure overheats the filler metal and oxidises it, producing an unsatisfactory joint.

It is preferable to select a flux which melts some 100degC below the melting point of the filler metal. Most manufacturers supply a range of fluxes which are suitable for use at various temperatures; low temperature fluxes have fluoride as an important constituent and those for higher temperature have borax. It is recommended that, where excessive fume is observed or where discomfort is experienced during brazing, both the brazing technique and type of flux being used should be investigated. Some of the more important toxic hazards are briefly described below.

Cadmium

This element is present in some filler metals, particularly those of the copper-silver-zinc type. Cadmium oxide fume and dust are very harmful and care is necessary to ensure that they are not inhaled. If the brazing procedure is correct very little cadmium fume will be present, but even so it is important for good local ventilation to be provided, except for very intermittent work in well-ventilated workshops. For production brazing local fume extraction is necessary, and for all work in confined spaces a face mask supplied with fresh air at a positive pressure should be worn.

If required, cadmium-free brazing alloys are available; as it is necessary to use more silver to keep a reasonably low melting point, they are more expensive. They are mainly intended for use in equipment which will be used in food preparation, as cadmium is also toxic if ingested in food or water.

Similar precautions must also be taken if cadmium is present as a plated surface on the parent metal. Alternatively, it may be acceptable to remove the plating mechanically, for example by turning or grinding (see Chapter 23).

Beryllium

Beryllium forms an important alloy with copper and is also a constituent of some magnesium and aluminium brazing alloys. It is highly toxic: serious illness and death have resulted from exposure to metallic beryllium and its compounds in the form of fume and dust, and stringent precautions should be taken to avoid contamination of the workshop atmosphere, and even of the atmosphere outside the factory. Precautions must be accompanied by regular analyses of the workshop air to determine the amount of beryllium present.

Any work on beryllium alloys must be regarded as an extremely hazardous operation which should not be started until suitable ventilation and other precautions are available and check analyses have been organised. In any event, the Health and Safety Executive in the UK, or other official inspectorate, should be consulted.

It may be necessary to supply employees with protective clothing to avoid contamination of their personal clothing, and close attention to personal hygiene should be encouraged by the provision of easily available washing facilities.

Zinc

Zinc may be present in the filler and parent metals, or as a coating, and the precautions outlined above for cadmium may be used as a guide. Though toxic, it is much less so than cadmium.

Fluxes

Fluxes are mixtures of inorganic chemical compounds formulated to give satisfactory results for specific purposes. Those for low temperature silver brazing are based on a combination of salts of sodium, potassium, boron, and fluorine. The fumes given off when these fluxes are heated may contain small quantities of hydrofluoric acid and boron trifluoride, which are likely to be irritant to the eyes, nose, throat, and respiratory passages. Consequently, local fume extraction should always be provided, with the possible exception of work of a very intermittent nature. Dermatitis can arise from skin contact with the fluxes and their fumes, so that precautions must be taken to avoid this hazard. A suitable barrier cream should be applied to the hands and forearms, and a protective ointment of the lanoline type (vaseline three parts, and lanoline one part) may be used on the face and inserted in the nostrils several times a day. Abrasions or breaks of the skin should be immediately covered with a waterproof adhesive dressing. Washing the hands must include thorough cleaning of the nails.

Fluxes for high temperature brazing have boric acid as a principal constituent and no significant ill-effects arise from their use.

Where fluxes have to be mixed into a paste with water, containers must be provided which will not be confused with teacups etc.

Similar fluxes to those mentioned above are used in flux baths. Dip brazing with fluorine-bearing fluxes should be done in baths provided with efficient exhaust ventilation.

The main use for salt bath brazing is on aluminium where nitrate salts are in common use. Apart from this there are some applications in low temperature silver brazing where the salts are composed of a mixture of sodium cyanide and sodium carbonate. When higher temperature salt bath brazing is used the salts will be based upon sodium carbonate.

Many of the salts are supplied under a proprietary label which does not disclose their exact chemical composition. However they are all designed to work over a given temperature range, and if this range is not exceeded little or no fume should be given off. Nevertheless, it is advisable to provide some form of local exhaust ventilation should the bath be used at the upper limit of its temperature range.

Local exhaust ventilation must be provided for cyanide salt baths. Cyanide salts and acids must not be stored together for their mixture will lead to the evolution of hydrogen cyanide, which is an extremely dangerous gas. Cyanide can be absorbed into the system through the skin, and the salts, on local contact, can give rise to dermatitis, so that the provision of protective clothing is advisable and employees should adopt a high standard of personal hygiene. These salts must not be introduced into the mouth and therefore meals and refreshment should not be taken at the workplace. Burns caused by splashes of molten cyanide must receive immediate first aid treatment. Special first aid treatment for cyanide poisoning should be readily available wherever cyanide salts are used; for further information refer to Chapter 27.

SURFACE PREPARATION AND CLEANING PROCEDURES

A clean, oxide-free surface is imperative to ensure a sound brazed joint, and all grease, oil, dirt, and oxides must be carefully removed from the filler and parent metals before brazing. Unsatisfactory surface conditions can be dealt with by mechanical means such as grinding, filing, scratch brushing, and various forms of machining, or by the use of a variety of chemical cleaning solutions. On occasions both mechanical and chemical cleaning will be necessary. Certain mechanical methods, such as grinding, will require the operator to protect his eyes with goggles and the dust produced to be removed by local exhaust ventilation.

It is not possible to give a detailed list of the many cleaning solutions in use for they vary considerably in their formulations, and some are supplied under a proprietary label which does not disclose their chemical composition. They can be roughly classified into two main groups: solvents, such as carbon tetrachloride, trichloroethylene, perchloroethylene, benzene and petroleum derivatives, and acids such as nitric, hydrochloric, hydrofluoric, and chromic. Certain chemical substances such as sodium nitrate, calcium or magnesium fluoride, and sodium dichromate can be added in small quantities to the acid solutions. Benzene and carbon tetrachloride are toxic, and benzene is highly flammable, so they should not be used as cleaning agents.

Adequate ventilation must always be provided in areas where cleaning operations are being conducted with the use of solvents. Chlorinated hydrocarbons, for instance trichloroethylene and perchloroethylene, are dissociated by heat and ultraviolet light to phosgene and other chemicals, which are irritant and harmful to the respiratory passages and to the lungs (see Chapter 23). Some solvents, such as petroleum derivatives, are inflammable and explosive and proper precautions should be taken against these risks. In the UK this may be required by law (Ref.16.3). Solvents should not be used as hand cleansers because they remove fat from the skin and render it sensitive to damage by other chemicals.

Unless acid solutions are very dilute it is advisable to install local exhaust ventilation over cleaning and pickling tanks to remove acid fumes, particularly if the process is hot or electrolytic (see Chapter 24).

Operators should be supplied with protective clothing, gloves, footwear, and a suitable barrier cream or lanoline ointment which can be applied to the skin of the hands, forearms, and face and, if necessary, inserted in the nostrils.

Hydrofluoric acid and its vapour are particularly corrosive to the skin, fingernails, and respiratory passages, and, if it is used in concentrated form, stringent precautions are required: in the UK, the Factory Inspectorate should be consulted. Its use in dilute form must be accompanied by the preventive measures outlined above, with adequate training and warning of operators.

It is dangerous to inhale the 'fumes' from concentrated nitric acid, therefore any spillage should be hosed away to a drain with large quantities of water by persons wearing efficient breathing apparatus. *The spillage should never be absorbed with a cotton mop, cloth, or sawdust.*

The use of caustic alkali solutions in tanks or otherwise for cleaning metal assemblies demands the provision of similar preventive measures to those recommended for acids. Assemblies that have been brazed, particularly if a flux has been used, may need cleaning, either by grit-blasting or by scrubbing by hand in a tank of hot water. The water will become contaminated with irritant flux residues, thus skin contact must be avoided.

The first aid treatment for splashes of acid or alkali on to the skin or into the eye is given in Chapter 27.

FURNACE ATMOSPHERES
The most common gases used in brazing furnaces are:

Hydrogen
Cracked ammonia
Burnt natural gas

These gases can give rise to accidents: there is a danger of the furnace combustion products contaminating the atmosphere of the workshop, and care should be taken to see that the blanket flame is present while the furnace is in operation. It is recommended that all products of combustion should be discharged to the outside atmosphere.

ELECTRICAL HAZARDS
Most of the electrical equipment is likely to be housed in a cabinet which should not be opened until the power has been switched off. With HF equipment the operator will not receive a harmful electric shock from touching the work coil, but he may suffer a HF burn which will be deep, painful, and slow to heal. Precautions should therefore be taken to avoid contact with, or close proximity to, the work coil.

Operators should not wear rings if they are able to bring their hands near the work coil while it is energised: rings will be heated just like the work.

PHYSICAL HAZARDS

The physical hazards encountered in brazing are somewhat similar to those experienced in welding; however, certain aspects merit particular consideration. In the operation of salt baths precautions must be taken to protect the operators against spatter, which can arise from a number of causes. The most common of these are the sudden escape of entrapped air from an assembly or from hollow tools, the rapid evaporation of entrapped moisture remaining as a result of inadequate preheating, and foreign bodies dropping into the bath. Particular care should be taken to ensure that workpieces are not dropped into the bath or left lying at the bottom.

When brazed aluminium assemblies are cleaned in tanks of caustic solution hydrogen may be evolved by chemical reaction, producing a spray or mist of caustic liquid. This will cause irritation, or worse, of the eyes, nose, throat, and skin if it is allowed to come into contact with them. The tanks should, therefore, be fitted with a system of local exhaust ventilation. Employees must be provided with goggles, protective clothing and footwear, and a suitable barrier cream for the skin.

An uncomfortable glare will be experienced during continual torch brazing; this can be avoided by wearing goggles fitted with filter glasses selected in accordance with Table 5.1 in Chapter 5.

BRAZE WELDING

Braze welding, a technique similar in some ways to both welding and brazing, is mostly used to join iron and steel components, but it can be applied satisfactorily to certain copper alloys.

Basically the process consists of heating the assembly with the cone of a gas flame as in welding, but using a lower melting point filler alloy, as in brazing, to provide the connection between the two parts. The filler metal is continuously fed into a pool of molten filler metal using a technique similar to gas welding, and the bonding of the joint results from wetting the unmelted surfaces and interdiffusion of the filler and parent metals. Many combinations of gases may be employed to produce the heating flame but the three most common are oxyacetylene, oxyhydrogen, and oxypropane. The process is such that dissimilar metals may be joined, provided the fusion temperature of both parent metals is above 950°C.

Many filler alloys are used but they are commonly rich in copper with small additions of nickel, silicon, tin, manganese, iron, aluminium, and lead. The application of flux is necessary which is likely to be of the borax type. The filler alloys are fed into an extremely hot flame and considerable fume may be evolved which should be controlled if necessary by the provision of local exhaust ventilation, as copper fume is a potent cause of 'metal fume fever' (see Chapters 23 and 24).

The operator can be continually looking at the hot cone of the flame, and incorrect use of his blowpipe may result in the filler metal splashing; he should, therefore, be supplied with protective clothing and goggles as recommended in Chapter 5.

CHAPTER 17

SOFT SOLDERING

PROCESSES
The processes normally used for soft soldering are:

Hand soldering with a soldering iron
Soldering with a blowlamp or torch
Dip soldering
Hot plate soldering
Induction soldering
Electrical resistance soldering
Furnace soldering
Standing wave soldering of printed circuit boards

The first method employs soldering irons heated by gas flames or electricity; the other methods operate on somewhat similar principles to those described under brazing. Hand and dip soldering involve the application of molten solder. In soldering with a blowlamp or torch, the solder is run into the joint which is heated with a flame, usually of the gas/compressed air type. With all the other methods, solder and flux are preplaced in or near the joint before heat is applied by the various methods which are characteristic for the individual processes.

TOXIC HAZARDS

Hazards from the filler metal
The soft solders in common use are alloys containing varying proportions of lead and tin. Lead poisoning can arise from the inhalation and ingestion of metallic lead and many of its compounds, particularly in the form of fume or dust. There is little risk to health in handling alloys in solid form, such as solder sticks or wires, but the inhalation or ingestion of finely divided solder, such as powder or filings, may constitute a hazard. The operation of sanding or filing down solder fillings on motor car bodies should be confined to a ventilated booth to prevent the spread of dust to other parts of the workshop, and the operator should be adequately protected by a helmet supplied with air at a positive pressure.

93

Food, drink, and tobacco must be kept away from the soldering bench and operators must wash their hands and arms before taking food. Ideally, food and refreshment should be consumed in a special room provided for this purpose.

Where clothing may become contaminated by lead dust it will be necessary to provide protective clothing which can be laundered under suitable conditions to prevent the spread of contamination. Lead is more toxic to young children than to adults, so the operators should preferably not take contaminated clothing home to be washed.

Dross skimmed from melting or dipping pots which contain molten solder should be handled with care so that it does not become airborne, and it should be collected and kept in containers with well-fitting lids: *keeping solder dross in sacks is dangerous. No dross should be allowed to collect on the floor.* In many instances the dross from solder pots will be mixed with residue from soldering flux; it is then in the form of a moist sludge and, although there is no risk of it becoming airborne, the substance is still toxic and should not come into contact with food or drink.

Whereas in normal circumstances soft solders are not liable to be heated to temperatures at which the formation of fume takes place, it is possible for solder to be spilled on to glowing coke or charcoal if the solder pot is heated on a brazier: an undesirable practice with any solder. Because the risk is particularly great with aluminium solders containing cadmium, this practice should not be used with these materials.

Hazards from the flux
Fluxes used in soft soldering fall into two main groups: active fluxes based on zinc chloride, and safety fluxes based on rosin. Zinc chloride is a corrosive substance and is toxic if taken internally; it should not be allowed to contaminate food or drink. It is a skin irritant and can cause dermatitis. The introduction of zinc chloride into the eye will give rise to a severe inflammation, and zinc chloride fumes may cause ulceration of the nasal passages. In most circumstances it is sufficient to insist upon good housekeeping practice and to instruct operators to avoid contaminating their skin or garments with the flux. Some operations which demand the use of large quantities of flux, such as the dip soldering of radiators, give rise to excessive evolution of flux droplets into the workshop atmosphere. In such instances a fume exhaust system should be installed close to the source of the flux vapour.

Whenever an active flux is used a barrier cream, supplied on medical advice, should be applied to the hands and forearms. The use of protective gloves should be discouraged because, once they become contaminated with zinc chloride on the inside, they are difficult to clean and promote rather than prevent dermatitis. Dermatitis is often caused, not through the action of the flux, but by operators using dangerous materials, such as solvents, to clean their hands. It is impossible to remove active flux from the hands with ordinary soap because it forms an insoluble compound with the flux. A degreasing agent, such as trichloroethylene or petrol, must not be used to wash the hands because it dries and cracks the skin, and then contact with an active soldering flux is painful, and can lead to ulceration of the skin and to severe dermatitis. Safe hand cleansers, which remove flux efficiently and do not lead to skin damage, must be used; medical advice may be needed to select a suitable formulation. Fluxes intended for soldering stainless steel and similar metals may contain hydrofluoric acid and should be treated with special care: observe the supplier's recommendations.

Active fluxes are eye irritants, and, where working conditions make it likely that flux will splash, safety goggles or screens should be used.

To avoid corrosion of the soldered job it is usually necessary to wash off the flux. This operation should also be checked for safety, such as safe wash water disposal. Establish a regular replenishment cycle for flux-contaminated water.

Safety fluxes generally contain rosin as a main constituent and often incorporate a volatile solvent. Rosin is not a dangerous substance but it can act as a skin irritant. In operations demanding the use of large quantities of flux the rosin fumes can be unpleasant, and local exhaust ventilation is recommended for such processes.

The common solvents present in rosin fluxes are methylated spirits or isopropyl alcohol. Constant inhalation of the concentrated vapour from these solvents, which is unlikely to arise in practice, could lead to mild symptoms of poisoning, i.e. lassitude, sleepiness, and headache. Where such a possibility exists, irritation of the nose, throat, and eyes from the rosin fumes would have already indicated the need for exhaust ventilation to remove the vapours from the atmosphere of the workroom.

It is impossible to remove rosin fluxes from the skin with soap, and a safe hand cleanser should be used, as previously recommended.

ELECTRICAL HAZARDS
Alternating current electric soldering irons of normal voltage must be earthed, unless they are double-insulated. Where it is desired to avoid earthing, as in working on certain live circuits in electronic equipment, a low voltage (12-24V) iron may be used with a suitable safety isolating transformer.

Where HF heating is used for soldering, the safety precautions previously described under Brazing in Chapter 16 should be observed.

PHYSICAL HAZARDS

Dip soldering entails the use of considerable quantities of molten solder, and articles and tools covered with flux have to be immersed slowly in the molten metal to give the flux time to boil off gradually. It is advisable to provide splash guards around the rim of the dipping bath, and protective goggles should be worn.

The flux boils off slowly on contact with the molten metal, but if a moist article is rapidly immersed in molten solder the moisture trapped on its surface evaporates, often with explosive force, and leads to the metal 'spitting'. Unless the bath is covered with flux, any article dipped in the molten metal should be dried by preheating and immersed slowly and carefully. Similarly, if molten solder is poured into a mould when a solder bath is emptied, the mould must be quite dry.

FIRE HAZARDS

Rosin fluxes contain volatile solvents which are inflammable. They do not constitute a high fire risk, *but all supplies must be stored in closed containers. Naked lights must be kept away from open flux baths.* If a flux bath should catch alight accidentally the flame can be extinguished by covering the container with a sheet of metal.

CHAPTER 18

THERMAL SPRAYING

Thermal spraying is widely used in industry both to reclaim worn parts and to deposit a surface coating of differing physical characteristics during production.

Only general principles are outlined here as it is not possible in this short account to deal with the many processes and materials used. It will therefore be particularly important to obtain further advice, for example from equipment suppliers or the UK Health and Safety Executive, if materials not mentioned in this Section are to be used, and there is doubt concerning their toxic or accident potentialities.

It is usually necessary for articles to be degreased and shot blasted before they are sprayed. General advice on the safe use of solvents such as trichloroethylene and perchloroethylene for degreasing purposes is given in Chapter 23. Sometimes pickling is used to remove a previous coating prior to fresh surface preparation; precautions should be taken to protect operatives against injury from acid and alkali splashes and from the inhalation of irritating vapours from the pickling tanks.

SURFACE PREPARATION BY BLASTING

Blasting is carried out with abrasive materials such as chilled iron grit, or steel or aluminium oxide grit; the use of sand or other substance containing free silica must be avoided. In the UK, legislation controls the permissible conditions of work (Ref.18.1).

In a factory, blasting operations should be carried out in a suitable enclosure or room to protect other personnel from injury and nearby machinery from damage. The 'blast room' should be provided with an efficient system of exhaust ventilation, preferably of the down-draught type. During the blasting operation abrasive material rebounds from the surface of the article with a high velocity, and consequently the operator must be given special protective clothing such as gloves, apron, and leggings. A helmet supplied with fresh air at a positive pressure is also necessary to protect the blaster from both flying particles and harmful dust, Fig.18.1.

Because of the friction between the finely divided particles of grit and the blasting hose and nozzle, it occasionally happens that discharges of static electricity take place. To protect the operator from this, it is advisable to earth the blasting hose and nozzle by means of a wire wound round them.

It is important that compressed air used in thermal spraying should be as free from oil and moisture as possible. Very moist or oily air, besides affecting the quality of the blasting and spraying, is not suitable for supply to items such as helmets and masks for breathing, as it will be unpleasant and may be toxic. The air from the compressor should pass to an after-cooler, thence to an air receiver for piston compressors, and a filter should be inserted in the airline just prior to the point of usage.

18.1 Correctly protected blaster

TOXIC HAZARDS

The greatest single application of the thermal spraying process is in coating ferrous metal with zinc or aluminium to prevent rusting. There are many important engineering uses of thermal spraying ranging from the reclamation of worn parts by building up with a suitable metal, to the application of numerous types of wear-control scheme. Spraying complete electrical circuits by utilising special stencils is an example of the more unusual uses of the process. Any metal, ceramic, plastic, resin, wax, etc. which melts without decomposition can be applied by thermal spraying techniques. Some of the other metals or alloys which are used are tin, copper, brass, bronze, cadmium, lead, and steel. A wide variety of ceramics, mixtures of alloys and intermetallic compounds or ceramics may also be sprayed using plasma and various types of combustion spraying apparatus.

It is probable that all metals can give rise to metal fume fever when they are inhaled in the form of fume or very fine dust. Consequently, it is always necessary to remove metal fume or dust from the breathing zone of the operator; this can be achieved by the installation of some form of local exhaust ventilation, preferably a ventilated enclosure or booth. Such a precaution may not be necessary for work in the open air. In confined spaces a respirator supplied with fresh air should be worn by the operator.

Lead
This metal is a highly poisonous substance which, after absorption, tends to accumulate in the body, eventually giving rise to harmful effects. In other words, a long period of exposure to small concentrations of lead will ultimately give the same result as a short period at higher concentration. Safety precautions must be stringent and the operators should be under continual medical supervision. The employees concerned must be protected by a helmet supplied with fresh air at a positive pressure. Spraying should take place in a ventilated enclosure, preferably isolated from other operations, so that the general atmosphere of a workshop is not polluted and other employees are not exposed to an unnecessary hazard. The hands and face must be washed before taking food and drink, and operators should have baths and change any impregnated clothing before leaving the factory. (See also Appendix 4.)

Cadmium
Cadmium is extremely toxic: precautions similar to those outlined for lead may be found satisfactory, but should be confirmed by measurements of airborne dust.

Plastics, resins, waxes
The powder process may be adopted to apply coatings of certain plastics, resins, and waxes by a simple modification of the spraying procedure. Some of these materials give rise to unpleasant fumes which are irritant to the eyes and respiratory passages; consequently the operator must be given adequate personal protection. The fumes must be removed by an efficient local exhaust ventilation system and discharged into the outside atmosphere at a high level.

PHYSICAL AND ELECTRICAL HAZARDS
A thermal spraying gun must never be directed towards any person. Protection of the eyes and skin of the operator should follow the requirements of the equivalent gas, arc, or plasma welding process described in earlier chapters. Where the arc or plasma stream is shielded from the direct view of the operator, a lighter filter glass will allow a clear view of the work.

Certain spraying guns, in particular arc and plasma guns, are excessively noisy in operation. In these circumstances a mechanised system may be used, if practicable, and the complete equipment housed in an acoustic booth, which can then be engineered to provide the necessary ventilation for fume protection also. If manual operation cannot be avoided, the operator and others exposed to the noise should use ear protection (see Chapter 8).

Arc and plasma processes involve similar dangers from electric shock to the equivalent welding processes and the safe practices prescribed in the appropriate chapters should be followed.

EXPLOSION AND FIRE HAZARDS

Certain metal dusts are liable to explode on ignition if they escape into a workplace in sufficient quantity. It is essential, therefore, to remove metal dust from spray booths or other confined spaces by adequate exhaust ventilation. All ducting for ventilation and dust collecting plant should be provided with blowout panels, and all equipment used for ventilation, including motors, fans, and pipes, should be electrically earthed. Explosion reliefs should not discharge into workrooms, therefore dust conveyors or ducting should be as short and strong as possible, or be sited outside workrooms. If it is found necessary to repair-weld or flame-cut such ducting, it should be thoroughly washed down and all metal dust removed, since the application of a torch to a ferrous base covered with a coating of certain dusts causes an exothermic reaction to take place, which damages the base and prevents proper welding from being carried out.

As far as the collection of metal dusts is concerned wet scrubbers are greatly preferable to dry-dust collectors. Bag or filter-type collectors should be sited outside the workrooms and be provided with explosion reliefs; the units should not be connected to any other processes for there is always the possibility of static electrical discharge resulting in a serious explosion. Cyclone-type collectors should also be located outside the building, preferably on the roof. It is well known that certain metal dusts, when partially wetted, are capable of spontaneous combustion, so the cyclones should be protected against the entry of moisture. *For good ventilation practice it is always desirable to consult properly qualified ventilating engineers.*

When cleaning out the booths, ductwork, and cyclones (and it is advisable to do this regularly to prevent excessive dust accumulation) the ventilation fans should be left running, all sources of ignition in the area should be eliminated, and non-sparking tools should be used for the cleaning and repair work. The structural steelwork in the spray shop should be cleaned down regularly, particularly on the top sides of the girders where most of the metallic dust accumulates, otherwise a damaged electrical cable or some other electrical defect may ignite the deposit and the ignition will travel where any inflammable dust is located.

A secondary explosion may follow if, for instance, a minor disturbance in the building causes vibration, shaking the dust lodging on the beams, in turn giving rise to a dust cloud of explosive concentrations within the workroom for a small instant in time. This cloud might be ignited and an explosion could then be generated on a massive scale. The efforts against this sort of occurrence must be directed towards the prevention of dust accumulations in the first place by:

(a) the proper design of buildings housing the hazard: for example, walls with smooth surfaces and where practicable the minimum number of ledges and obstructions on which dust accumulations are possible
(b) the modification of existing buildings, including fittings to prevent such dust accumulations: for example, boxing in girders and the removal of unnecessary fittings

Where an explosion hazard can arise, all possible sources of ignition should be excluded or effectively enclosed (Refs 18.2 and 18.3).

Aluminium dust must not be collected in the same receptacle as iron dust from blasting or spraying. If aluminium dust catches fire, it should be extinguished with dry sand; *water must not be used.*

A thermal spray must not be directed towards a container which holds, or has held, combustible materials, and spraying should not be carried out in close vicinity to these materials.

The greatest risk of fire in thermal spraying comes from the use of inflammable gases such as acetylene, propane, butane, etc., and the safe practices outlined in Chapter 2 should be followed.

To a great extent, however, the danger of fire has been removed from the thermal spraying industry, by special installations and by equipping the 'danger areas' with the most efficient firefighting devices (see Chapter 26). These techniques include the use of bulk liquid oxygen and propane supplies sited outside the spray shop.

Where natural gas from supply mains is compressed the installation must be approved by the supply authority, who will require nonreturn valves and other safeguards against the formation of an explosive mixture in their mains.

Where quantities of propane are stored, staff should be trained in the use of whatever extinguishers and other firefighting equipment are provided.

THERMAL SPRAYING EQUIPMENT
The flame spraying guns in common use are light, easy to manipulate, and

perfectly safe in operation; their main differences are in the form in which the metal or other material to be sprayed is supplied, e.g. molten metal, wire, or powder. The metal is heated by burning a mixture of oxygen and a combustible gas which produces an annular flame around the central orifice of the nozzle. An outer ring of high velocity air has the dual purpose of keeping the flame short and relatively cool, and also picking up the hot metal and projecting particles on to the job. Wire or powder is fed into the heating zone where it is melted by the flame. A gas mixer is incorporated in the gun to ensure satisfactory mingling of the gases, and also to act as a flashback arrestor should the gun backfire.

If a thermal spraying gun blows out no attempt should be made to relight it before the reason for the failure has been found and the cause eliminated. Care should be exercised in the cleaning and lubrication of guns: oil should not be allowed to enter the gas passages and mixing chamber, and only special lubricants recommended by the manufacturers should be used for valves and moving parts. The rubber hose used to convey oxygen to the gun must be clean, free from oil, and of the correct blue colour. For the gas hose, orange hose with a nitrile rubber inner lining should be used for propane or other LPGs, and red rubber hose for others such as acetylene, in either case with a left-hand thread screw nipple. Oxygen and compressed air lines should have right-hand thread nipples of differing sizes.

Acetylene gas must be employed only in guns which are specially designed for use with it, since a high issuing velocity of the oxyacetylene mixture from the gun nozzle is required to prevent the flame from burning back to the mixer, because of the very high rate of flame propagation of such a mixture. Guns for use with acetylene should be clearly marked as such, and close attention should be paid to the manufacturers' instructions. Pipes, fittings, manifolds, etc. made of unalloyed copper must never be used with acetylene gas because of the possibility of the formation of explosive cuprous acetylide.

Other thermal spraying systems make use of the electric arc or arc plasma as the heat source and safe handling requirements will be similar to those of the corresponding welding processes.

RECOVERY OF OVERSPRAY
A certain amount of the coating metal is carried away as overspray; this may be sufficiently valuable as scrap to be worth collecting by means of a recovery system. If the powder dust being recovered is toxic, any operators who are exposed to it when unloading filters etc. must be adequately protected.

CHAPTER 19

WELDING AND FLAME SPRAYING PLASTICS

WELDING PLASTICS

Plastics are well established as engineering materials and, although the great bulk of this material is moulded into its final shape, there are a number of applications where it is required to fabricate the end product from such forms as sheet, bars, rod, or pipe, or where mouldings forming the case of a mechanism such as a gearbox must be assembled by welding. Such processes being used are:

Hot gas welding
Heated tool welding
Friction welding
High frequency welding
Heat sealing (mainly of films and foils)

Almost all thermoplastics may be welded, for example:

Rigid polyvinylchloride (PVC)
Plasticised polyvinylchloride
Polyethylene, low density and high density
Polypropylene
Penton (chlorinated polyether, no longer generally available)
Perspex (polymethylmethacrylate)
Nylon (polyamide)
Polystyrene
Polyvinylidene chloride
Ethylcellulose
Cellulose acetate
Cellulose acetate butyrate
Polychlorotrifluoroethylene
Cellulose nitrate
Rubber hydrochloride
Polyvinylalcohol

Hot gas welding

The fundamental technique of hot gas welding is similar for most thermoplastic materials. The sections to be joined are cleaned and slanted, and a filler rod of similar or special composition is used. The weld is made by heating the weld area and the filler rod in the hot gas issuing from the nozzle of the welding torch. When the edges to be welded have reached a bonding temperature, they are kept together under pressure until cool. The bonding temperature for most thermoplastics is in the region of 280° to $370^{\circ}C$.

Equipment: that used to gas weld plastics can be roughly subdivided into welding torches which are heated by commercial gases and those which are heated by electricity. The former contain a heater coil which is heated by a heater gas, and the latter have a heating element; both heating mechanisms are devised to allow the passage of a welding gas. Nitrogen is a commonly used welding gas, particularly in this country, but certain plastics can be welded with compressed air, or carbon dioxide.

Gas-heated torches: these use a fuel gas (acetylene or propane) and air for heating.

The hazards encountered in the operation of gas-heated torches are the same as those found with other appliances heated by such gases. The design of welding torches has now reached a stage where the danger of blowback and explosion through the ignition of unburnt gas has been reduced to negligible proportions. The risk of explosion of cylinder-stored acetylene gas is fully described in Chapter 2.

To avoid toxic or explosion hazards it is essential that the flame of the burner does not extinguish after lighting.

A serious hazard to be avoided when welding in enclosed spaces is that of asphyxiation, due to the accumulation of welding or heating gas or the products of combustion such as carbon dioxide. Care should therefore be taken to ensure that sufficient artificial ventilation is provided, for instance when welding inside closed containers (see Chapter 24).

The most obvious hazard presented by a gas-heated torch is that of burns caused by accidentally touching its heated parts. In general there is no necessity to wear gloves because the temperatures are much lower than in the oxyacetylene welding of metals, and there is little likelihood of the operator suffering a severe burn.

One type of welding torch consists of a coil through which air is circulated, and external to which is an acetylene flame which heats this air in its passage through the coil. To conserve the heat of the acetylene flame the coil is enclosed in a thin steel container with an opening at the top through which the end of the heating coil projects, forming the torch for the application of the heated air to the plastic material to be welded. This container is a thin, double-walled steel case, the annular space being tightly packed with asbestos. It is essential that there are ample vent holes in the outer cylinder of the container, as a serious accident has resulted by the generation of steam pressure in the annulus owing to the asbestos being damp.

When welding is finished and the heater gas has been turned off the welding gas should be permitted to flow for at least five minutes to cool the heating coil. Similarly, when the heater gas flame is ignited it should not run for too long a period without a flow of welding gas.

Electrically heated torches: these can be connected directly to the 240V mains, in which event it is essential that an earth wire be fitted (or a double-insulated construction used) to avoid a shock if the heater windings make accidental contact with the torch casing. Many factories permit the use of portable electrical appliances only with a voltage lower than 60V, such as a 110V centre-tapped supply. Welding torches adapted to operate from such low voltage supplies are available and the danger of electric shock is very much reduced with such instruments.

Another hazard which must be avoided with the use of electrically heated torches is the same as that for gas-heated torches, i.e. burns through accidental contact. Provided reasonable care is taken this does not constitute a great danger and in most conditions the use of protective insulated gloves is not essential.

It is recommended that the welding gas be allowed to flow for approximately five minutes after switching off the current so that the element can cool.

The attention of users of electrically heated torches is drawn to the need for regular inspection and maintenance to avoid electric shock hazards.

The usual precautions regarding ventilation should be taken when welding in a confined space, noting in particular the risk of asphyxiation from a build up of welding gas.

Heated tool welding
Plastics materials can be welded by the application of heated tools of various types, for instance an electrical strip heater, hot plate, ring, bar, or soldering iron, heat being transmitted either by direct contact or by radiation. Heated tools, the metal surfaces of which may be nickel-plated, steel, or solid aluminium, have a large variety of shapes and are most commonly used for the jointing of pipelines

and to weld flanges to pipelines. In such circumstances the interior of the flange and the exterior surface of the pipelines are heated by metallic appliances and subsequently, after removal of the appliances, brought together under pressure so that the liquefied plastic surfaces come into contact and, on consolidation, form a proper joint.

Tools of this type can be heated by gas burners as well as by electricity. Temperatures of operation depend somewhat on the particular type of plastic material to be joined and would in most instances not exceed 200° to $250^{\circ}C$. The main hazard in this type of welding is burns from accidental contact with heated equipment.

High frequency welding equipment
A typical machine for the HF welding of plastics consists of a press which is provided with a fixed and a movable system of electrodes between which the article to be welded is clamped at the time of operation. Power is supplied to the electrodes from a HF generator and frequencies of the order of 20-150MHz are most commonly used to weld thermoplastic materials such as plasticised PVC and cellulose acetate.

The principal hazard of HF welding is one of contact by the operator with a live electrode, as a result of which he is likely to suffer a HF burn. Such a burn can be very painful as it tends to penetrate deeply into the tissues, affecting tendons and nerves; subsequently, healing may take a long time. High frequency welding machines are usually supplied with adequate guards to prevent access to the live parts of the system. Care is needed to maintain this protection where the user builds or modifies tooling, or when covers are removed for maintenance (Ref. 19.1).

TOXIC HAZARDS
The majority of thermoplastic materials which are welded for engineering applications are highly polymerised and are comparatively inert; therefore, they are unlikely to give rise to any serious health hazards. Plastics which are unstable or which are subject to thermal degradation at welding temperatures cannot be satisfactorily welded, and those having constituents which volatilise during welding are unlikely to produce sound welds. Nevertheless, before welding any unfamiliar plastic material it is always advisable to consider the possibility of a toxic hazard from gases and fumes, so that adequate precautions can be instituted if these are necessary. Information may be sought from the manufacturers or suppliers of the plastic. This particularly applies where accidental overheating of the plastic is likely to occur: for instance, the overheating of PVC results in the production of small quantities of hydrogen chloride gas, although the concentrations are unlikely to reach hazardous proportions in an adequately ventilated workshop.

At temperatures in excess of 250°C polytetrafluoroethylene (PTFE) decomposes with the evolution of toxic fluorine compounds which, if inhaled, produce symptoms similar to those experienced during an attack of metal fume fever. This material must not be welded unless exhaust ventilation is applied as near to the weld as possible. In addition, smoking in the workshop must be prohibited and smoking materials should not be taken into the workshop to avoid their contamination with PTFE dust and fume. Employees should be provided with overalls and good washing facilities adjacent to their workplace.

FLAME SPRAYING PLASTICS

The only plastics which may be successfully flame sprayed are castor base polyamide, polyethylene polysulphide, epoxy resins, and chlorinated polyether (no longer generally available).

In the main, the equipment for flame spraying plastics is identical to that for flame spraying metals. Equally, the surface to be sprayed is normally prepared in a similar manner, that is to say, by shot blasting. Most problems on health and safety while flame spraying plastics will already have been covered by the precautions suggested for the handling of fuel gases and oxygen in earlier chapters. However, certain additional possibilities should be borne in mind:

1 Molten thermoplastics may adhere to the skin and cause deep burns
2 On oxidation, chlorinated polyether, polyethylene polysulphide, epoxy resins, and cellulose acetate butyrate give off pungent irritating vapours with a disagreeable smell
3 Plastics containing chlorine may give off harmful acidic vapours when heated in the presence of air

Personnel will not be affected by obnoxious and harmful vapours if flame spraying is carried out in an adequately ventilated enclosure.

None of the thermoplastics which can be readily flame sprayed presents a serious fire risk if the equipment manufacturers' instructions for the use of the equipment are followed correctly. Finely divided polythene and epoxy resins can ignite in the flame of a spray gun if the rate at which the powder is fed to the gun is too high. Under these conditions a flame may be thrown for several metres. Prompt action, in first shutting off the powder supply then the gas supply, will eliminate the danger.

Some of the plastic dusts used can constitute a Class 1 dust explosion hazard. Overspray and the exhaust equipment provided for its disposal may also give rise to an explosion hazard. The precautions which should be followed are those outlined in Chapter 18.

CHAPTER 20

RADIOGRAPHIC INSPECTION

The examination of welds by X-rays and ionising radiations from sealed radioactive sources is common practice. The most common types of isotope used for radiography are cobalt 60, iridium 192, and caesium 137. Ionising radiations can neither be seen nor felt and undue exposure to them may be seriously harmful. Such effects may be delayed and vary widely: from a transient reddening of the skin appearing within a week or two to, say, leukaemia, a fatal blood disease, which occurs after perhaps five or ten years. It is essential, therefore, for precautions to be taken to limit personal exposure to ionising radiations to an amount less than that recommended internationally as 'permissible'. The UK regulations require precautions to be taken to ensure that the dose to the operator and to anyone else is minimised, and shall not exceed certain prescribed values. It should be understood that radiation exceeding the permissible limits by several times would be needed to cause a medically observable effect, such as a change in the composition of the blood.

The legal requirements for the safe use of ionising radiations emitted by sealed sources or X-ray sets in the UK, covering the major part of industrial radiography, are laid down in Ref.20.1 and explained in Ref.20.2. The other nondestructive testing (NDT) technique which requires special safety precautions is penetrant testing, which is dealt with in Chapter 23.

MEASUREMENT OF RADIATION
New SI units for ionising radiation were introduced in 1978:

Absorbed dose: 1 gray (abbreviated 1Gy) = 1J/kg = 100rad
Dose equivalent: 1 sievert (abbreviated 1Sv) = 1J/kg = 100rem (for the usual radiographic sources, the gray and the sievert are numerically equivalent)
Activity: 1 becquerel (abbreviated 1Bq) = 1 disintegration/sec
3.7×10^{10}Bq = 1 curie (1Ci)
Exposure (same as before): 1 röntgen (abbreviated 1R) = 2.58×10^{-4} coulomb/kg

SUPERVISION AND TRAINING
Precautions must be taken to protect both the radiographer and those incidentally

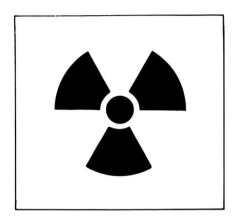

20.1 *Ionising radiation warning symbol. Preferred colour is black on chrome yellow background (Ref.20.3)*

exposed, for example, welders working nearby. The UK regulations require medical examination of, and the wearing of dosemeters by, 'exposed workers', that is, those who work with ionising radiations and who are not at all times outside areas in which the dose rate exceeds 50mSv (5rems) per year. This is normally interpreted as 7.5µSv (0.75mrad) per hour. For the general public (that is all except exposed workers), the dose limit for whole-body radiation is 5mSv (0.5rem) in a year. There are special regulations for pregnant women, women of reproductive capacity and persons under eighteen years old.

First of all, a 'competent person' with adequate knowledge must be appointed to take charge of radiation safety, and he must have sufficient authority to ensure that unsafe practices are eliminated. Next, radiographers must have formal instruction and examination on the hazards to which they may be exposed. Furthermore, it is important for welders to be made aware that ionising radiations can be dangerous. They must be taught to recognise radioactive source containers, X-ray equipment, and warning signs (Fig.20.1), to respect barriers and demarcation tapes, and never to tamper with sources or source containers.

The regulations require a number of registers and records, such as of dosemeter readings, to be kept in a prescribed form.

EQUIPMENT
When purchasing equipment attention should be given to safety requirements. Mechanical design and the degree of shielding are factors of importance both when the source is exposed and unexposed. In X-ray sets, safe electrical installation is essential to avoid any risk of shock. Dose rate meters should be frequently checked and calibrated.

SPECIAL PRECAUTIONS
A walled enclosure should be built when radiography is regularly carried on at a particular location and, if it is economically practicable, on open construction sites. Such an enclosure has especially thick walls to reduce the exterior dose

rate to, in general, below 7.5µSv (0.75mrad) in air per hour. In certain circumstances it may be possible to position the source so that the beam can be directed only on to one wall; if so, economies can be made by reducing the screening required on the other walls. The regulations specify the use of interlocks and warning signals. Safety devices should be carefully considered before installation so that they are both effective and foolproof. The exposure of a small source inside local shielding within the walled enclosure must not invalidate the main warning system.

On open or temporary sites or in *in situ* radiography in the welding shop there should be similar warning systems, using lights and, when appropriate, hooters. Physical barriers should be used to prevent approach to the source closer than the 7.5µSv (0.75mrad) in air per hour contour: or orange coloured demarcation tapes, supplemented by ample warning notices, may be used.

The removal of hand tools and personal property belonging to the factory workers from inside the demarcated zone around the source eliminates a temptation to illicit entry.

Remote source handling and control devices should be employed to minimise the dose to the operator. **No source should be touched with the bare hand.** During setting-up a dummy capsule, painted white, should take the place of an exposed source. The working beam should be directed away from occupied locations, and radiations outside the beam, including scatter, reduced by the application of recognised techniques.

The barrier or tape is normally positioned at the 7.5µSv (0.75mrad) in air per hour contour. The positioning must be checked by measurement when the source is exposed. If tape is used, visual supervision of the entire perimeter throughout the source exposure period may be impracticable, particularly if the radiation zone extends over more than one floor level. A watch should, however, be kept from outside the tape on the exposed source itself. When a source is within its container and unexposed an inner perimeter may be found convenient, but it must be clearly indicated by warning notices, and must not be allowed to coexist with the outer boundary.

STORAGE
Stores of radioactive sources should be sited away from occupied areas, shielded as necessary, labelled with warning signs, and kept locked. The sources should be in separate shielded compartments, or in their own containers, so that they may be removed individually without causing unnecessary exposure. Large sources to be used wholly or mainly in walled enclosures should be stored there. On open or temporary sites storage in pits should be considered, care being taken over humidity problems. When the larger sources are removed the container shielding should be checked for defects with a dose rate meter as well as visually.

EMERGENCY
Emergency plans should be made to cover liaison with fire brigades besides action in the event of a mishap to the source. Special clothing and sensitive contamination measuring instruments should be readily available.

All radiographers working with isotope sources must receive training and updating of information regarding the procedures to be followed if a source is lost, mislaid, or cannot be returned to its container.

SUMMARY
Appoint a Radiation Safety Officer, the 'competent person' with knowledge and authority. Train radiographers in safe procedures and welders to recognise radiation hazards; each individual to accept responsibility for the avoidance of undue radiation exposure to himself and to other workers.

Use radiation meters continually and check their functioning frequently.

Aim to do rather more than the regulations require.

FURTHER INFORMATION
Makers of radiographic equipment will advise on precautions to be taken in its use; training in safety procedures will form an integral part of most training courses for radiographers. Procedures for construction sites are detailed in Ref.20.2. The UK authority is the National Radiological Protection Board, Harwell, Didcot, Berkshire, which operates the dosemeter system. At the time of writing, a revision of the regulations is proposed incorporating the new units.

CHAPTER 21

MECHANICAL HAZARDS

The mechanical hazards presented in welding and cutting are common to most engineering work, but there is a change of emphasis: particular attention should be given to the following points.

SAFE PLATFORMS
When working where a fall to a lower level is possible, a safe working platform should be provided. Open edges should be protected by handrails and toeboards; where appropriate, a safety belt should be worn.

OBSTRUCTIONS
Working areas should be kept free from obstructions as far as possible. This is particularly important where a welder or cutter may have to move while working, as his eye protection filter will restrict his vision.

The absence of accumulations of rubbish, slag, etc. makes it easier to avoid damage to hoses and cables, and to see any damage which does occur.

LIFTING

Mechanical
A number of unsafe situations can arise during the lifting of work. Wire ropes may be damaged by hot work or sharp edges, or even by welding current passing through them (see Chapter 7). Work may have been built up to a weight in excess of the safe working load of the lifting gear or work positioner, or its centre of gravity may be in an unsafe position.

Tack welds or untested welds, such as those holding temporary lifting lugs, may part when they bear stress on lifting.

Props welded between work and floor may not have been completely removed.

Manual
Persons lifting appreciable loads in the course of their work should receive

training in how to do this without injury. Basically the back is kept straight, upright, so that the vertebrae of the spine are loaded only in compression. This is done by picking up items from the floor from a squatting position, rather than by leaning over the load.

MANIPULATORS AND POSITIONERS
Rotary tables or rollers used to position work so that all welds can be made in the best position, for example flat or horizontal-vertical, present a number of hazards which have been overlooked in the past by welders.

A safe working load, maximum work dimensions, and allowable out-of-balance load should be established to avoid overstressing the equipment. Where work is moved during welding, the return path for welding current must be considered: if a cable is used, will it coil up safely as work proceeds, or will an assistant be needed to guide it? If a cable is not used, the return current must not be allowed to pass via the bearings and damage them, but through a proper slip-ring and brush, usually provided on equipment intended for this duty. Manipulators should be securely fastened to a sound foundation, or a large out-of-balance load may cause them to tip over suddenly during welding.

As with any motor-driven equipment, the welder should have ready access to an emergency stop button.

Where circular work such as drums or pipes is rotated on rolls, suitable precautions should be taken as required to detect and rectify any tendency for work to creep along its axis of rotation. Work with holes in the outer diameter, or projections such as stub pipes, may foul rollers or fixed plant during rotation.

Particular care is needed to avoid starting with a safe piece of work, which is then built up to a weight or size exceeding the capabilities of the equipment in use.

WIRE FEED UNITS
Wire feed units, used particularly in metal-arc gas-shielded welding and in mechanised welding and surfacing by a range of processes, are often capable of exerting enough force to drive the sharp end of the wire into the operator's hand. Operators should not place their hand over the gun and pull the trigger to check the gas flow: they should use the gas purge facility and/or place their hand clear of the wire.

TESTING

Pressure
Generally, pressure testing in which the components under test are filled with water is safer than testing with air, which stores much energy when compressed.

cidents have occurred, mainly when there are air pockets in a Reference 21.1 gives a series of rules.

from testpieces are mechanically tested there are a number of minor. They should be guarded against by proper training of those who carry out the tests, especially where they are not laboratory personnel but welding staff. Apart from the obvious need to keep clear of the moving parts of the testing machine when in use, makeshift arrangements for bend tests can allow packing or rollers to be ejected at high speed.

CUTTING
The portions of cut work to be severed should be supported as necessary to avoid injury or damage from falling pieces.

GRINDING
Portable grinding tools must be adequately maintained for safe operation. Electric tools should be checked for earth lead continuity or double-insulation as appropriate (see Chapter 7). Air tools must be used only from an appropriate airline with a sound hose (never from an oxygen supply, see Chapter 4). Wheels must be correctly chosen to suit the speed of the tool and correctly mounted; this is required in the UK by Ref.21.2.

PROTECTION

Eyes
Workers carrying out deslagging, chipping, or grinding should use appropriate eye protection. This is required in the UK by Ref.21.3; protectors to the relevant standard (Ref.21.4) are approved.

Personal
In most welding shops, safety footwear will provide valuable extra protection. Where processes involving substantial amounts of molten metal, such as electroslag or thermit welding, are in use, a pattern designed for use under foundry conditions will be more appropriate (Refs 21.5 and 21.6). On many sites safety helmets are required; some patterns are available which combine or can be worn with a welding helmet.

GUARDING MACHINERY
Some potentially hazardous parts of machinery, such as motor drives or work feed rollers, may be made safe by guards; Ref.21.7 lays down guidelines.

CHAPTER 22

MEASUREMENT AND ASSESSMENT OF FUME

The methods of providing personal protection against damage to the eyes, electric shock, and other injuries which can occur if proper precautions are not taken, have been outlined in previous chapters.

This chapter and the two which follow it deal with the problem of the airborne pollutants arising from welding and allied processes. For brevity all airborne pollutants will be referred to as 'fume', the word being taken to include airborne dust, particulate fume, and pollutant gases. Normally 'dust' is taken to mean particles of 0.5μm diameter or over, and 'fume', particles of less than 0.5μm diameter. This chapter outlines some considerations in the measurement and assessment of fume; Chapter 23 describes the sources of fume as related to the welding, cutting, or other process being carried out; Chapter 24 discusses the various methods available to control exposure of the worker.

Air pollution stemming from fumes arising from welding or associated processes can take two forms. The first is particulate fume or smoke, discrete solid bodies suspended in the air; the size which may be breathed in is up to 10μm diameter, Fig.22.1. The second is pollutant gases which mix freely with the air and therefore cannot be removed by a membrane type filter. (Charcoal filters can, however, absorb or adsorb gases to a limited extent). A possible exception is ozone, which reacts chemically with almost all filter materials.

MEASUREMENT OF FUME CONCENTRATIONS
Fume measurement in practice falls into three parts. In the first, particulate fume is collected on a filter by drawing the polluted air through it; by weighing the filter before and after exposure the mass of fume per unit volume of air is determined, the usual units being milligrams per cubic metre (mg/m^3). In the second, chemical analysis of the deposit on the filter enables the concentrations of individual elements and compounds to be found. In the third, which will often be carried out simultaneously with the first, gases are measured by a chemical reaction in a gas detector tube or a special-purpose analytical

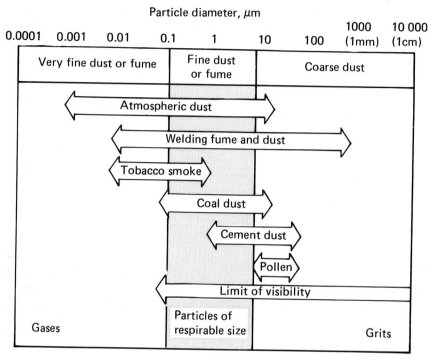

22.1 Sizes of particles

instrument; the amount present is expressed as a concentration in parts per million (ppm).

A British Standard is in the course of preparation on measurement of welding fume (Ref.22.1).

THRESHOLD LIMIT VALUES
The measured concentrations of particulate and gaseous pollutants are compared to the Threshold Limit Values (TLVs) which are officially published as a guide to the maximum concentrations to which personnel may be exposed. A lower TLV figure indicates a more harmful substance. These limits are estimated by medical experts on the evidence of case histories of industrial disease and the results of animal experiments, and are reviewed as more information becomes available and revised as necessary. Lists of these TLVs are published annually by the American Society of Occupational Hygienists, and republished in the UK by the Health and Safety Executive (HSE) (Ref.22.2).

The latest list may be taken as a legal guideline to the maximum level to which workers should be exposed. However, there is usually a considerable variation in measurements obtained under apparently identical conditions, and occasional excursions to higher concentrations are permitted for most substances, so one or two test results on their own do not immediately prove acceptable or unacceptable working conditions. Even if levels are below the TLVs, however, this may not meet the requirements of the HSE that exposure to pollutants should be reduced to the lowest level which is reasonably practicable.

As TLVs are subject to regular revision, and are published on condition that extracts from the full list of values are not to be republished, they are not quoted in this book. Those qualified to make fume measurements will normally have a copy of the complete tables, or will know the up-to-date TLVs for the pollutants they are equipped to measure.

Some pollutants will have an immediate effect if inhaled in excessive concentrations, but others may take effect only after many years of exposure; the effects may be temporary or permanent. Individuals vary in their tolerance to pollutants and the ease and effectiveness of treatment available will vary. Factors such as the above are all considered in fixing the TLV, but the figure stated does not report this information and so should not be taken to indicate the relative severity of the effects of different pollutants.

Further information
See Refs 22.3 and 22.4. Organisations offering a service for the measurement of welding fume are listed in Appendix 2, Part 2.

CHAPTER 23

SOURCES OF FUME

This chapter details the sources of fume, with notes on those procedures which are available to reduce or remove it at source.

The welder may be exposed to particulate fume and gases arising from:

Surface treatment, and penetrants used in NDT
Macroetching
Action of the heat source on the parent metal
Action of the heat source on the surface coating of the parent metal
Action of the heat source on the surrounding air
Action of the heat source on the consumables (filler, flux, or shielding gas)
Internal combustion engines

SURFACE TREATMENT
It is often necessary to remove dirt, grease, paint, rust, and scale from metal surfaces prior to welding, brazing, soldering, and thermal spraying so that work can be carried out without defect, and to grind work after welding and cutting.

Abrasive and chemical methods are available for surface preparation. Abrasive methods include grit and shot blasting, mechanical grinding, hand filing, and the manual use of steel wool and a wire brush. Safety precautions for grit and shot blasting are outlined in Chapter 18. Protection for the eyes against flying particles must be provided when powered grinders are in use, and may be advisable with manual abrasion (see Chapter 21).

Where dust is released in excessive quantities, or is toxic, local exhaust ventilation or respiratory protection will be required. Where parent metal has a surface coating, such as galvanising, which would produce undesirable fume if welded, it may be acceptable to remove this coating by grinding, taking suitable precautions (but for cadmium plating, see below).

Chemical methods involve the use of solvents, for instance trichloroethylene, carbon tetrachloride, benzene, toluene, white spirits, and solvent naphtha, and of acid, alkali, and detergent solutions. Solvent cleaning is carried out by means of hand swabs or by immersion of the workpieces into tanks. Carbon tetrachloride and benzene are such poisonous substances that their use, even in small quantities, should be avoided. Commercial toluene (toluol) has in the past been found to contain up to 15% of benzene, which makes it a dangerous cleaning agent, hence a purer distillate with less than 1% benzene is now standard (Ref.23.1); both toluene and benzene are also highly flammable. In addition to their inherent toxic properties the chlorinated aliphatic hydrocarbons, such as carbon tetrachloride, trichloroethylene, and perchloroethylene, can be decomposed by heat or by ultraviolet radiation, with the evolution of phosgene and other toxic gas. Their use in association with welding and allied processes should, therefore, be the subject of special care. Stabilised 1.1.1 trichloroethane, to Ref.23.2, is the least toxic of the chlorinated solvents in common industrial use, but is still liable to be decomposed by heat or ultraviolet radiation. This solvent should not be allowed to remain in contact with aluminium for any length of time, as it may promote decomposition which can proceed explosively.

Degreasing operations should preferably be carried out in a workshop separated from and nonadjacent to the welding location. If such an arrangement is not possible the general ventilation should always be such that the direction of air flow is away from the welding area to the cleaning bay, and not vice versa.

Where tanks are used for degreasing they should be so located that air currents from doors and windows do not cause dispersal of vapours into the workshop atmosphere. Tank covers should be used whenever possible, and workpieces should be placed into and removed from the tank at low speed to minimise displacement of vapour and 'drag-out' of degreasing liquid. Exhaust systems in which extraction takes place through slots along the sides of the tank are recommended. All traces of the solvent should be removed from the workpieces by a sufficient period of drying, preferably by a hot air blast, before they are welded.

Manual application of degreasing fluids should be carried out only under conditions of good general ventilation, and preferably in a fume cupboard or booth-type of enclosure fitted with exhaust ventilation so that the vapours are drawn away from the breathing zone of the operator. A safe container for the disposal of used swabs should be provided: on no account should they be placed in the operator's pocket. There is also the alternative of using a solvent which does not decompose, but all practical choices will be flammable, with a low flash point, so care is still needed, though this at least is a straightforward hazard. On balance, the best choice appears to be acetone, with no fume problem, but a pungent smell to warn of its presence; it is highly flammable.

Pickling in cold or hot acid solutions, to remove rust, scale, and oxides, can give rise to air contamination by gases and acid mists, and exhaust ventilation through slots at the rear of the tank may be necessary. The air volumes necessary for the slot exhausts will depend upon the nature of the operation and the location of the tanks, and will range from 15m^3/min per m^2 of tank surface for degreasing, to 75m^3/min per m^2 or more for pickling or acid cleaning. Disposal of used solvents or pickling baths can present difficulties as they cannot normally be allowed to enter the public drainage system; it may be possible to arrange for used solvents to be returned to the supplier or a specialist processor for distillation and re-use.

If welders have to strip lagging from work before welding, for example from pipes, asbestos may release harmful fibres, requiring suitable protection measures. Asbestos blankets should not be used in heat treatment or preheating operations.

Nondestructive testing by penetrants
In the penetrant methods of NDT a dye is applied to the job by spray, swab, or immersion. If there are any surface cracks in the job the dye enters them. Surplus dye is then cleaned from the surface, usually by wiping, and the final stage is to detect the dye trapped in the cracks, either by applying a chalky white powder developer into which the dye soaks to give a brightly coloured stain, or by viewing in ultraviolet light which makes the dye fluoresce. Moderate care is needed to avoid exposure of the eyes to ultraviolet light, but the intensity is much less than in arc welding. The solvents and aerosol propellants can present similar problems to those used for degreasing, and any precautions recommended by the manufacturer of such fluids should be carefully observed.

MACROETCHING
To examine test welds a macrosection may be prepared. A cross-section of the weld is cut, and one of its faces polished. Then the polished face is etched heavily to reveal the crystalline structure of the metal in examination with the naked eye or a low-power hand magnifier.

The foremost safety requirement is suitable eye protection from the etching solution with goggles or a clear screen: facilities for washing the eyes (see Chapter 27) should be available as close as possible in the event of an accident.

Etching solutions should not be allowed to come in contact with the skin: tongs should be used to hold specimens and it may be advisable to wear rubber or plastic gloves. Special care is needed for solutions containing hydrofluoric acid or strong alkalis such as caustic soda, as detailed in Chapter 16 under the heading 'Surface preparation and cleaning procedures'.

Fumes produced during etching are unpleasant and usually harmful, so a well-ventilated place is desirable for this work. Ideally, etching should take place in a laboratory-type fume cupboard, but at the very least there should be a large deep sink with an adequate water supply. A notice should clearly state the hazards of the various etchants, the safe working procedures to be employed, and the measures to be taken in the event of an emergency.

As some etching solutions will react violently if carelessly diluted they should be mixed by competent persons. Different etches should not be mixed together as a violent reaction may again occur. Safe disposal facilities for spent solutions are also required.

ACTION OF THE HEAT SOURCE ON THE PARENT METAL
As the parent metal is raised to its melting point in welding it is likely to generate fumes, usually containing the oxides of the metal. The fume from steels or aluminium is not usually troublesome, but other more toxic materials such as copper or its alloys may produce fume levels exceeding their TLVs unless ventilation is good. Table 23.1 will help to identify the constituents of alloys which are important in the formation of fume and Table 23.2 indicates likely pollutants: they should be read in conjunction with the text of this chapter. For processes other than welding or cutting, the parent metal does not usually contribute significantly to fume, for example in brazing copper.

Specific pollutants
Turning now to specific pollutants, the inhalation of excess copper or zinc fumes may give rise to metal fume fever (see Chapter 27).

Irritant manganese fumes have been reported in isolated instances when welding high manganese (12-14%) steels.

Cadmium and beryllium are extremely poisonous: the effects of cadmium are described in the next Section, and beryllium causes specific changes in the lungs (berylliosis) which are usually fatal. Great care is therefore needed before attempting to weld them (see also Chapter 16). There is, however, little information available as to the degree of toxic hazard presented by the welding of copper-beryllium alloys, as used for nonsparking tools.

Lead melts at such a low temperature (327ºC) that it can be welded only with a low heat input process, such as oxygas or pulsed tungsten-arc gas-shielded, which generates little fume in normal operation. (See also Appendix 4.)

Cutting
In cutting, substantial amounts of parent metal are released as oxide or vapour.

Table 23.1 Constituents of alloys

MATERIAL (*trade mark or trade name)	CONSTITUENTS (those in *italics* contribute significantly to toxic fume)
Steels	
Mild steel	Iron
Stainless steels	Iron, chromium, nickel, molybdenum, cobalt
High yield steel	Iron, *manganese*
Aluminium alloys	Aluminium, *manganese, zinc* (some contain copper, but are not usually welded)
Magnesium alloys	Magnesium, aluminium, *zinc, manganese, thorium*
Alloys with special names	
Aluminium bronze	Aluminium, *copper*
Brass	*Zinc, copper*
Cupronickel	*Nickel, copper*
German silver	*Copper, zinc, nickel*
Gunmetal	*Copper, zinc, lead,* tin
*Incoloy	*Nickel,* iron, *chromium, copper*
*Inconel	*Nickel,* iron, *chromium*
Manganese bronze	*Copper, manganese*
*Monel	*Nickel, copper*
*Nimonic	*Nickel, chromium, cobalt*
Phosphor bronze	*Copper, lead,* tin
Soft solder	*Lead,* tin
Alloys for casting	
Aluminium-based alloys	Aluminium, and some contain *copper*
Copper-based alloys	*Copper,* and some contain *lead*

Aluminium bronze
  Brass
  Gunmetal
  Phosphor bronze
  Nickel alloy castings
	details as for 'Alloys with special names' above

Table 23.2 Significant pollutant constituents.

Constituents with TLVs higher than the nuisance particulate level (10mg/m^3) at the time of compilation (1980), such as aluminium and iron, have been omitted. The ratings are based on continuous use of the process in an open shop

	Particulate	Nuisance particulate	Barium, Ba	Beryllium, Be	Cadmium, Cd	Chromium, Cr	Cobalt, Co	Copper, Cu	Fluoride, F	Lead, Pb	Manganese, Mn	Nickel, Ni	Silver, Ag	Thorium, Th	Gases	Carbon monoxide, CO	Oxides of nitrogen, NO$_x$	Oxygen enrichment, O$_2$	Ozone, O$_3$
Welding																			
Oxygas																			
Mild steel	1															1	1	1	
MMA																			
Mild steel	2	(2)							1							1	1		1
Manganese steel	2	(2)							1		2					1	1		1
Stainless steel	2	(2)				2	(2)		1			2				1	1		1
Metal-arc gas-shielded																			
Mild steel: CO$_2$ gas	2							1△								2			1
Ar gas	2							1△								1			1
Stainless steel	2					2						2				1			2
Manganese steel	2										2					1			1
Aluminium	2																		2
Copper	2			(3)				3											1
Mild steel:																			
flux-cored wire	2	(2)							1	1						1			1
Tungsten-arc																			
gas-shielded																			
Mild steel	1													X					2
Stainless steel	1					1						1		X					2
Aluminium	1													X					2
Copper	2			(3)				2						X					2
Nickel	2											2		X					2
Submerged-arc																			
Steels	1																		
Cutting and gouging																			
Oxygas																			
Mild steel	2									2*						1	2	2	
Air-arc																			
Mild steel	2							2		2*						1	1		2
Oxyarc																			
Mild steel	2									2*						1	1	2	2
Plasma																			
Mild steel	2									2*						1	1		2
Aluminium	2																1		2
Brazing																			
All processes	2				(2)							2							
Soldering																			
All processes	2									1									
Preheating																			
Flame																2	1		

Key:
- ( ) occasional occurrence only
- X possible significant radiation hazard when grinding electrodes only
- △ from electrode wire antirust plating
- * lead-painted or lead-coated sreel
- 1 significant amounts of fume, but not usually exceeding TLV
- 2 precautions recommended, such as local ventilation
- 3 potential danger from quantity and toxicity

A substantial reduction in atmospheric pollution can be achieved by placing the work just above, or partially immersed in, a shallow water tray; this technique is mainly applicable to machine oxygas or plasma cutting.

ACTION OF THE HEAT SOURCE ON THE SURFACE COATING OF THE PARENT METAL

Plating

Metal is commonly plated with another metal by dipping, such as in galvanising by immersion in molten zinc, or by electroplating, such as with chromium or nickel. The heat of the welding arc or other heat source tends to cause vaporisation or oxidation of the plating, and the fumes are sometimes harmful.

Zinc plating, as galvanising, will often give enough fume to cause metal fume fever as described above.

Cadmium plating is used as an anticorrosion treatment, particularly on light engineering products such as office machinery and electronic connectors. In appearance it is indistinguishable from zinc plating, and both are frequently given a passivating treatment which imparts a golden sheen. Even a small amount of fume may be fatal: it causes metal fume fever as above, in some cases followed by acute inflammation of the lungs (see Chapter 27). In view of its toxicity, conventional machining or a chemical treatment are preferred to grinding for its removal.

Lead plating is used on steel for corrosion resistance, particularly as 'terne plate' for motor vehicle fuel tanks. These are usually resistance welded in production, when fumes may be kept under control by ventilation, but difficulties may be encountered in repair. Of course, the precautions outlined in Chapter 9 may also be required to prevent an explosion. Lead poisoning can be detected reliably by blood lead measurement.

Chromium is often used as a decorative plating, but the layer thickness is probably too thin to release a significant amount of fume. Thicker hard chrome deposits applied to reduce wear may be cut or welded; it is likely that a significant quantity of chromium may be present in the fume. (The hazards of chromium in other industrial applications in the form of an acid mist are likely to be greater than those of the oxides formed in welding.)

Most commercial 'chromium' plating uses undercoats of nickel and copper, which make up the bulk of the total thickness. In adverse circumstances these could give rise to excessive fume. At the time of writing the hazards of nickel and copper fumes are under investigation.

Table 23.3 Contaminants likely to be formed by heating paint and plastics coatings

Elements present in resin	Chemical classification of resin	Possible products of pyrolysis
Carbon, hydrogen, and possibly oxygen	Resin and derivatives Natural drying oils Cellulose derivatives Alkyd resins Epoxy resins (uncured) Phenol-formaldehyde resins Polystyrene Acrylic resins Natural and synthetic rubbers	Carbon monoxide Aldehydes (particularly formaldehyde, acrolein, and unsaturated aldehydes) Carboxylic acids Phenols Unsaturated hydrocarbons Monomers, e.g. from polystyrene and acrylic resins
Carbon, hydrogen, nitrogen, and possibly oxygen	Amine-cured epoxy resins Melamine resins Urea-formaldehyde resins Polvinyl pyridine or pyrrolidine Polyamides Isocyanate (polyurethanes) Nitrocellulose derivatives	As above, but also various nitrogen-containing compounds, including nitrogen oxides, hydrogen cyanide, isocyanates
Carbon, hydrogen, and possibly halogens, sulphur, and nitrogen	Polyvinyl halides Halogenated rubbers PTFE and other fluorinated polymers Thiourea derivatives Sulphonamide resins Sulphochlorinated compounds	As above, but also halogenated compounds. These may be particularly toxic when fluorine is present Hydrogen halides Carbonyl chloride (phosgene) Hydrogen sulphide Sulphur dioxide

Paint

Industrial paints may contain metals or their compounds for specific protective purposes: for example zinc metal or lead oxide (red or white lead) in an antirust paint, or mercury compounds in marine antifouling treatments. Most paints will have complex natural or synthetic organic binders which can decompose to form a wide range of fumes when heated. A specific, widespread, and serious danger is that of lead fumes when welding or cutting lead-painted steelwork, especially in a confined space.

Plastic coating

Plastic and synthetic resins are now commonly used in coating materials for the protection of metal surfaces and as insulating layers between metals. Burning and welding such coated metals invariably gives rise to organic decomposition products, many of which are known to be hazardous to health. All coated surfaces must be treated as potential sources of noxious fumes, particularly where the nature of the coating is not known (as is usually the situation during maintenance and repair work).

If the nature of the coating materials is known it is sometimes possible to predict the decomposition products. In view of the great number of synthetic resins in use, and the variety of formulations that are possible, no comprehensive list of air contaminants can be given. Each case must be reviewed separately, and even then the full answer can be obtained only after actual tests on the coating in question. From past experience it is possible to make some generalisations about the air contaminants likely to arise from certain classes of coating material, and these are tabulated as Table 23.3; detailed information may be found in Ref.23.3.

In addition to the organic materials mentioned above, the coating material may contain a variety of metals and inorganic materials such as zinc, zinc chromate, iron oxide, aluminium, etc., all of which may create extra hazards. The complexity of modern coating materials is such that several different resins may be used in the same formulation, as well as novel materials not mentioned above.

The quantity of noxious compound evolved is dependent on a number of factors, including:

Formulation of coating
Area of coating burnt
Thickness of coating
Temperature of pyrolysis
Presence of oxygen excess or deficiency

The last two factors may also affect the nature of the pyrolysis products: for example, the absence of oxygen will favour the formation of carbon monoxide and unsaturated compounds.

It should be remembered that, during welding, the reverse side of the metal will also be raised to a high temperature, which may give rise to fumes in other compartments or rooms.

The increasing use of plastic foams or expanded plastics, e.g. polystyrene or polyurethane foams, for thermal or acoustic insulation can present an additional hazard during maintenance and breaking-up operations. Because of the presence of large amounts of organic material, cutting or burning the insulated section could produce dangerous concentrations of air contaminants.

Precautions when welding etc. on coated material
Unless any fumes generated are known to be harmless, which can be established only by tests carried out under conditions comparable with those applying in practice, it will be necessary to take precautions by way of adequate ventilation and/or personal protection, or by local removal of the coating other than by burning. This latter will also ensure that the coating does not interfere with the joining process, and that paint etc. is not baked on and rendered difficult to remove. It will in most instances be desirable to remove any overheated coating, and replace it over the weld itself to gain the benefits of corrosion protection.

As explained above, grinding or other abrasive removal will usually be suitable except for cadmium plating. The coating should be removed from all surfaces of the work within 25-100mm of the weld. It will often be clear whether removal has extended for an adequate distance to be effective, but for arc welding 25-50mm is suggested as a guide, work at higher currents (around 200A and above) requiring the greater distance, and for flame processes 50-100mm, depending again on circumstances.

ACTION OF THE HEAT SOURCE ON THE SURROUNDING AIR

Ozone
The major components of normal air are approximately 78% nitrogen, 1% argon, and 21% oxygen. When exposed to ultraviolet light from the arc the oxygen atoms can rearrange themselves into ozone:

$3O_2 \longrightarrow 2O_3$
diatomic   triatomic
oxygen   ozone

Ozone is chemically very active and toxic and contact with most solids will cause it to revert to oxygen; this makes it difficult to measure. A conventional filter will therefore partially remove ozone from polluted air, unlike any other pollutant gas likely to be encountered in welding which would need an absorption filter.

The processes producing most ozone are the gas-shielded welding of aluminium and its alloys, during which precautions are advisable against exposure of welders to excess ozone: for example extra ventilation may be provided, or dust respirators or other personal protection (see Chapter 24) may be used.

It is also claimed that ozone formation may be inhibited by adding 0.05% of nitric oxide (NO) to the argon shielding gas. This mixture has been patented and is commercially available.

All ozone found in a welding shop may not stem from welding. It has been observed in the course of research at The Welding Institute that ozone may be found in appreciable quantities near fluorescent lamps, presumably arising from their emission of some ultraviolet radiation.

Oxides of nitrogen
Where flames are used under conditions of restricted ventilation, or for preheating, nitrogen and oxygen from the air may combine to form various oxides of nitrogen such as nitric oxide (NO) and nitrogen dioxide (NO_2); they are usually quoted simply as oxides of nitrogen, with the abbreviation NO_x. They form nitric and other acids in contact with moisture and, as they are a potentially serious hazard if present in excess, adequate protective measures are essential.

ACTION OF THE HEAT ON THE CONSUMABLES

Filler
In welding, any filler rod will normally be of similar composition to the parent metal and so should present no new hazard, but the filler will often be heated to a higher temperature than the parent metal, and so may produce rather greater amounts of fume. There are two common exceptions to this: the first, MMA welding, is discussed in the following Section. In the second, copper applied as rust-inhibiting plating to low carbon steel filler rods may be vaporised and cause appreciable pollution. The process in which this is most likely to occur, metal-arc gas-shielded welding, has, however, been shown to exceed the TLV for copper only where ventilation is insufficient to prevent the total fume level also being excessive, so no special precautions should normally be required.

In brazing and soldering, on the other hand, the composition of the filler will be markedly different from that of the parent metal, and so extra precautions will be needed if it contains potentially toxic constituents such as cadmium, beryllium, copper, and lead. These examples have been described in more detail in Chapters 16 and 17.

Flux
As with filler materials, the fluxes used in brazing and soldering, and the possible toxic fumes which may be evolved from them, have been noted in Chapters 16 and 17.

In MMA welding the electrode core wire carries a covering whose major function is that of a flux. Electrodes are classified according to the major constituent of their covering. The first class is 'cellulosic', containing the organic material cellulose, a hydrocarbon which decomposes in the heat of the arc to form such gases as carbon dioxide, water vapour, and hydrogen, with a fair amount of dust as a byproduct. The total fume tends to be greater from the cellulosic type of electrode than from others.

The second, 'rutile', contains rutile, an impure mineral form of titanium dioxide. It is the normal general-purpose electrode used for most run-of-the mill jobs on low carbon steel. It tends to produce less fume than other types.

The third, 'basic', refers to the metallurgically basic behaviour of the covering; a major constituent is calcium fluoride in the form of fluorspar. It is used where the covering must not contribute hydrogen to the weld, to provide a low hydrogen or controlled hydrogen electrode, as used to weld steels containing appreciable amounts of carbon and manganese to strengthen them. It tends to produce fumes containing fluorine compounds which are irritant and may be toxic.

In addition to the above three main constituents iron powder may be added to the covering by the electrode manufacturer to increase welding speed. It tends to increase total fume by increasing the amount of iron oxide. Alternatively, alloying elements may be added to the covering of a low carbon steel wire so that a high strength or stainless steel, or other desired alloy, may be deposited. These elements may be toxic and may be present in the fume.

Considering also the wide range of sizes of electrode in common use, from those with a core wire diameter of 1.25mm up to those of 6.4mm, and the range of current over which each size of electrode may be used, it will be seen that a range of variables needs to be considered, and that any measurements should be carried out under conditions corresponding closely to those under which the job will be done.

Nevertheless an attempt has been made to measure the total amount of fume emitted by a given size and type of electrode at maximum current, and to use this information to assign each size and type of electrode to a fume category. This project originated in Scandinavia, and possible further development may result in international agreement on a universally acceptable procedure.

The MMA process as used in the past is known to have exposed workers to fume containing much iron oxide. This settles in the lungs and, though producing no symptoms of ill-health, is visible in normal chest radiographs; this condition is known as 'siderosis' or 'welder's lung'. This builds up over a period of years if exposure is continued, but is gradually dispersed if it is stopped. With modern electrodes, working practices, and ventilation standards, this condition is becoming less common.

In automatic metal-arc welding an electrode with a flux covering and retaining wires is continuously fed into the arc by a mechanical drive. Though the system would appear similar to MMA welding in many ways, the amount of fume generated is usually much greater, in which a large part must be played by the higher currents and continuous operation without stops for electrode changes.

In metal-arc gas-shielded welding flux is not normally used, but in some applications in the welding of thick steel, or the surfacing of steel, with a hard deposit, the electrode may take the form of a tube containing a powdered flux, drawn down to a diameter as small as 1.2mm. Welding with current types of flux-cored wire may be carried out with or without shielding gas, and produces a great deal of fume; with or without shielding gas; current types of flux-cored wire tend to produce considerable amounts of fume but future developments may result in a reduction in fume evolution. Some specific flux-cored wires have in the past been found to evolve fume containing barium.

The flux used in submerged-arc welding usually melts without producing any significant amount of fume.

One instance in which a particularly active flux is required, and which is likely to emit toxic fumes, is in the obsolescent welding of aluminium by the gas or MMA processes, therefore the consumable manufacturer's recommendations as to ventilation should be sought and followed. Currently, gas-shielded welding offers much better results and much less fume evolution.

Tungsten-arc gas-shielded welding
In the tungsten-arc gas-shielded or TIG process the electrode is normally described as nonconsumable, being made of tungsten which does not readily melt or evaporate, though gradual vaporisation of the end of the electrode causes

it to be very slowly consumed. Tungsten concentrations arising from this are not significant, but when using DC a 'thoriated' electrode is used to stabilise the arc: 1 or 2% of thoria (thorium oxide) is alloyed with the tungsten. Thorium is radioactive, emitting weak alpha-rays at a very low rate as it has a long half-life (10^{10} years). For radioactive materials the criterion in the UK is not airborne concentration TLVs but total accumulation of the substance in the tissues of persons exposed. It is currently accepted that the normal operation of thoriated electrodes presents no hazard. However, from time to time the end of the electrode must be reground. A welder grinding electrodes for his own use will probably not be overexposed to dust, but an operator grinding batches for several welders may breathe in excess thorium. Local ventilation is suggested, with provision for the safe disposal of polluted air or filtered dust.

Gases
In flame processes, complete combustion leads to end products of water vapour and carbon dioxide. The main effect of these will be asphyxiating, that is depriving the worker of breathable air, but carbon dioxide has a significant narcotic effect, producing feelings of well-being and tiredness which discourage action to remedy the situation. The combustion products will be hot and will rise away from the work zone, hence they are likely to build up only where ventilation is restricted, as in work in a confined space. Carbon monoxide may be formed as a result of incomplete combustion of the fuel gas. This is aggravated if sufficient oxygen for full combustion is not available, as in oxygas flames adjusted to a reducing condition, or a fuel gas burning in a space where the air supply is restricted (as might be found in preheating). Even in small concentrations it combines with the haemoglobin of the blood, preventing it transporting oxygen from the lungs to the rest of the body. The main symptom of carbon monoxide poisoning is a bright red coloration of the skin.

In gas-shielded welding the inert shielding gas is released to the atmosphere and, if allowed to build up, carries the danger of asphyxiation. This applies whether the shielding gas is carbon dioxide, argon, helium, or nitrogen; carbon dioxide, as explained above, is more toxic because of its narcotic effect. Few problems are found in normal work in the open, but gas can build up in tanks and other confined spaces, especially if it leaks from equipment during a work break. Argon and carbon dioxide are heavier than air and so may accumulate even in a tank with an open top. Carbon dioxide is known to decompose partially at the temperature of the arc, forming some carbon monoxide, but the process will be reversed as the temperature falls, and again there is little difficulty in normal work.

INTERNAL COMBUSTION ENGINES
When arc welding is to be carried out where a mains electricity supply is not

available, or is insufficient, a generator driven by an internal combustion engine will be used. The exhaust from the engine will be mainly inert gases (nitrogen, water vapour, and carbon dioxide) but with an appreciable proportion of carbon monoxide, especially where the fuel is petrol (gasoline) rather than diesel. If possible, the engine should be located in the open air, or where the fumes will dissipate harmlessly without being blown towards the welder or other workers. If it must be located indoors, a leak-tight extension pipe can be used to lead the gas to the outside atmosphere.

CHAPTER 24

VENTILATION AND FUME PROTECTION

This chapter first discusses the various working conditions, then ventilation and personal fume protection, concluding with recommended practice. The main concern is with the control of air pollution as a direct toxic hazard to health by general or local ventilation; this will also be effective where there is a risk of oxygen enrichment, or of fire or explosion, but it should be noted that protection will be reduced where gases are recirculated through filters which will not remove oxygen or inflammable vapours, and that personal protection, such as an air-supply helmet or dust respirator, will not be effective against these risks. Further advice is available from the sources listed under 'Further information' at the end of the chapter.

WORKING CONDITIONS

In the open air
This represents the situation in which least supplementary ventilation is required to reduce any given hazard. Exposure may be minimised if it is possible for the welder to stand upwind of the weld, so that the wind carries the fume away from him.

In the open shop
The great majority of welding operations are carried out in an enclosed workshop. Usually this will be of sufficient height to allow convection currents from the welding process to carry fume well above head height. However, unless fume is efficiently extracted it may form a layer in the upper part of the building, or even at an intermediate height. This may expose to excessive fumes those working above floor level, for example drivers of overhead cranes or welders on tall jobs.

In a semi-confined space
Where the floor area of a large workshop is divided into small welding bays natural air circulation may be obstructed to a significant extent. Where it is necessary to work inside a structure large enough to allow a reasonable amount

of room for the worker, the fume from welding operations will not disperse freely. In these conditions, and in those where work is carried out in a small closed room or booth, fume extraction equipment will normally be required.

In a confined space
Sometimes a weld must be made in a location where ventilation is severely restricted, for example, in a tank which has only a manhole opening, inside a casting with a bore only just large enough to allow the welder to get in, or a long way into a large structure. Fume extraction and/or other means of protection is nearly always essential in such circumstances.

VENTILATION

General ventilation
General ventilation is here taken to refer to a situation where air is extracted from the general volume of a workshop, and its flow is not confined artificially to the neighbourhood of welding work. To be effective, enough air must be extracted to reduce fume concentrations to an acceptable level, air flow should be well distributed with no stagnant pockets, and fresh air must be supplied (warmed or cooled if necessary) to replace that extracted. Though it may be expensive to install and run, it need cause little further interference with work.

Reference 24.1 quotes requirements in terms of 'Air changes per hour' (ach). The figure for factories is 6-10ach, but this is not intended as a guide for workshops where fume is a problem and it is not really a good basis: if the workshop height is increased, the ventilation required to maintain the same ach figure is increased correspondingly, whereas common experience suggests that less will often suffice.

An American Standard (Ref.24.2) has a more useful approach, i.e. mechanical ventilation is called for if volume per welder is less than $284m^3$, ceiling height is less than 5m, or work is in a confined or poorly ventilated space where cross ventilation is obstructed. The suggested normal minimum rate is $57m^3$/min per welder with increases for large or flux-cored electrodes. These flow rates will not necessarily meet current TLVs.

When planning a system, exhaust and inlet air flow should be considered, Fig.24.1. Ideally, it should be such that fumes will be drawn away from every welder's breathing zone, but in practice it will often be necessary to accept that much of the air will contain some fume, and it is possible only to limit this to a safe value, rather than aim for completely unpolluted air.

Welder position
The natural tendency for a welder is to stand and bend over the work placed on

a welding bench. As the hot fume-laden air from the arc rises vertically it enters his breathing zone. Thus, if he adopts a posture so that his head is no longer directly above the arc, his exposure to fume will be reduced. In practice, the easiest way is for him to work seated, if possible.

Local extraction
Local extraction is here taken to mean exhaust ventilation where the major part of the airflow is confined to the immediate neighbourhood of the weld. It will normally comprise an extractor nozzle, air hose, fan unit, and discharge system; the last either discharges fume-laden air to the outside atmosphere or filters and recirculates it.

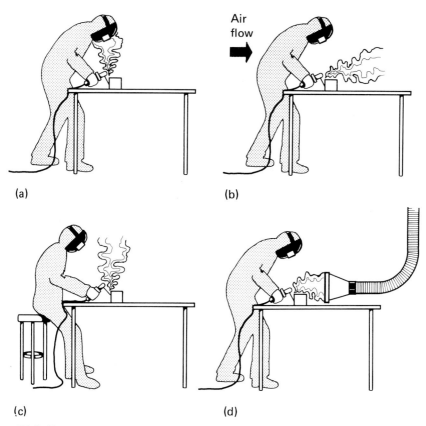

24.1 *Fume protection: (a) excessive fume exposure, (b) general ventilation, (c) improvement by welder position, and (d) local extraction*

The extractor nozzle should be easily positioned close to the weld: for example some commercial units have magnetic clamps. It should produce an air velocity of 0.5-1m/sec over as long a section of the weld as is practicable. Lower velocities will not generally be effective as fume tends to rise at about 1m/sec in convection currents, and higher velocities may remove the gas shield on which the gas-shielded processes particularly, and much MMA welding, rely. It should allow the welder a clear view of, and unobstructed access to, the work. The hose should be flexible and resistant to spatter, and the fan unit should be readily portable and reasonably quiet in operation.

Unfortunately, the above requirements tend to be mutually exclusive: a large nozzle extracting large quantities of air will need a large hose and a powerful fan for instance. The limited coverage obtainable from current designs means that the extractor nozzle must frequently be moved along the work as commercial nozzles are no longer than 0.5m.

Mechanised welding can be more easily, effectively, and reliably ventilated by such means as attaching extractor nozzles to the machine, so that any fume which does escape is less likely to enter a worker's breathing zone. Therefore, if all other measures are ineffective in the control of fume from a manual welding process, it may be worth considering mechanisation.

Fume extractor gun

In the metal-arc gas-shielded process, or variants using a 'self-shielded' wire which requires no separate gas supply, the normal gun may be fitted with an extractor nozzle surrounding the normal contact tip and nozzle assembly, Fig.24.2. When correctly designed and operated, effective fume removal is achieved without extra tasks for the operator. However, the bulk and weight of the gun is increased, and the extractor hose will be a substantial encumbrance, so that such a system may be difficult to use on other than jobs offering easy access to the joint.

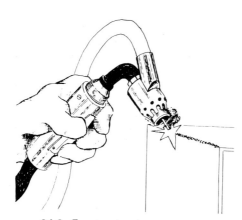

24.2 Fume extractor gun

Such a device has been marketed both as a complete gun with fume extractor built on, and as a clip-on attachment suitable for a range of makes and types of gun. The main application to date seems to have been in connection with flux-cored wire, with which it is able to remove sufficient fume to give entirely acceptable working conditions.

Though very effective in the removal of particulate fume in the arc region these devices may not remove the toxic gases, such as ozone and oxides of nitrogen, produced by ultraviolet light from the arc acting on the atmosphere at some distance from the arc. Indeed, the removal of particulate fume may increase the radiated ultraviolet, and hence increase toxic gas formation. Additional general ventilation may be needed to remove these gases.

External discharge
Extracted polluted air may be discharged to the atmosphere outside the workshop. Though simple in principle, there are three potential problems. Firstly, especially with local ventilation, an extensive hose or trunking system may be neeeded in large buildings. Secondly, especially with general ventilation, the make-up air drawn in to replace that extracted may need heating to raise it to a reasonable temperature in the British winter climate, or cooling in the Tropics. Thirdly, it is possible that it would be unacceptable to pollute the outside atmosphere, depending on the siting of the factory. All these difficulties can be overcome if it is possible to clean the air by removing pollutants and then recirculate it.

Filters
Where, as in most local or general extraction, the major pollutant is particulate fume, a filter will effectively purify extracted air for recirculation, but it will not deal with asphyxiant or pollutant gases, oxygen enrichment, or explosive hazards.

To recirculate local extraction, commercial portable units are available combining a fan and disposable filter, but they are considerably bulkier than a fan unit alone. Where a filter is used, regular cleaning and/or replacement will be needed; the man who carries out this operation may need protection from toxic dust, and used filters must be safely disposed of. The administration and financial implications should be taken into account when the system is selected.

Currently, no filter appears to be available which would be suitable to recirculate general ventilation in a welding shop with its heavy dust and fume load.

Electrostatic precipitator
In an electrostatic precipitator air is passed between two flat metal plates, typically spaced by 10mm, connected to a high voltage ($\sim$15kV) low current power supply. Any particle passing between the plates becomes charged, and is then attracted to one plate or the other by electrostatic forces. The plates are long enough in the direction of air flow to intercept all relevant particles before the air stream can carry them clear of the plates. In practice a series of plates,

connected alternately to positive and negative poles of the high voltage supply, increases the capacity. When the plates have collected so much dust that a discharge can occur between them, the filter plates are removed from the unit and brushed or washed down. (Although the power supply is of such limited current as to be safe to touch, interlocks are provided.) It has been alleged that these devices may produce substantial quantities of ozone, but there appears to be no documented evidence implicating current commercial products. It is possible that past designs had sharp edges to the high voltage parts which would be likely to encourage corona discharge, a potent ozone producer.

Electrostatic precipitators may be applied to recirculate air from either local extraction or general ventilation systems.

Fresh air supply
In combination with extractors, particularly in confined spaces, clean filtered fresh air may be fed via suitable trunking to the general area of work. The fans for this duty are similar to those for extraction but usually of somewhat greater capacity. An extractor draws in air from all directions, but the air supply is more directional and should be arranged to blow towards the welder and carry fumes to any extraction vents.

PERSONAL FUME PROTECTION

Dust respirator
If the fume from welding operations cannot be removed from the atmosphere before it reaches the welder, an alternative is to filter it from the air that he breathes with a dust respirator.

The half-mask respirator usually used in welding fits over nose and mouth, and carries a filter for air entering and a valve for air leaving, Fig.24.3. An alternative design has no valve: the complete mask is disposable.

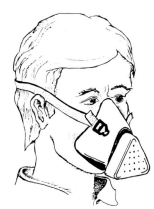

24.3 Dust respirator

A good fit is essential to prevent leakage: beards may have an adverse effect. The filter must be effective in removing the welding or other fume, and not obstruct breathing to the extent that it causes the welder appreciable discomfort. It will not remove gases, with the possible exception of ozone. The respirator must fit comfortably under the headshield if one is in use.

Canister or cartridge respirators relying on a chemical reaction, or on absorption, to remove specific pollutants are not generally used in welding operations: there is a risk that welding fume, unless effectively removed by the prefilter, would block the intended reaction, thus reducing the protection against the specific pollutant.

As with all personal protection the welder himself will be protected, but ancillary workers need special consideration. Further information is given in Ref.24.3.

Air-fed welding helmet
A welding helmet in which clean air is fed to the welder's breathing zone combines protection against a number of hazards, Fig.24.4. A correctly designed helmet will distribute the air comfortably, and will avoid any venturi effect which might tend to draw in fume-laden air at the edges. Cool air will improve comfort in hot surroundings.

Air may be fed by a hose, which should be appropriately secured to the welder's body in case it is caught up or pulled. If a compressed-air line is used it should not feed the mask directly: oil, water, etc. normally found in airlines must be filtered out, or it is usually found more practical to use a device where fresh air is drawn in, using compressed air only to power it.

Helmet with fan and filter
A welding helmet has recently been introduced which incorporates a fan unit to draw in air and deliver it through a filter to the welder's breathing zone; note again that normal dust filters will not remove gases. British Standard terminology for this equipment (Ref.24.3) is a 'positive pressure powered dust respirator'.

Breathing apparatus
Breathing apparatus, which supplies air to the breathing zone in such a way that air cannot be drawn in from the outside, permits work in irrespirable or poisonous (but not flammable or explosive) atmospheres, Fig.24.5. Where the

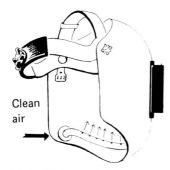

24.4 Air-fed welding helmet

24.5 Breathing apparatus

atmosphere is sufficiently toxic to put life at risk if equipment should fail, great care and adequate rescue arrangements are essential; it may even be considered unacceptably hazardous to combine such operations with welding. Where breathing apparatus is used in atmospheres immediately dangerous to life, competent supervision and trained personnel are essential, a safe system of work, such as a 'permit-to-work', must be adopted, and all necessary steps must be taken to provide a rescue facility.

The air may be supplied from a backpack cylinder or through a hose.

Further information on the equipment is given in Ref.24.4.

RECOMMENDED PRACTICE

General recommendations
In view of the wide range of welding and allied processes now in use, the numbers of different parent metals encountered in common practice, and the variations in natural workplace ventilation, general recommendations will apply to only a few of the many possible combinations of the above factors. Table 24.1 provides a few general recommendations, covering the most used processes and materials.

Procedure in individual cases
1. Establish the constituents of parent metal and consumables, or obtain a written assurance from the manufacturer/supplier as to their safety with or without special precautions
2. For processes listed in Table 24.1 use the minimum precautions where parent metals are mild steel or aluminium, or where they and the consumables contain no especially toxic materials
3. Where substantially toxic materials such as copper, nickel, or zinc are involved, follow Table 24.1. If work will be prolonged, take expert advice on the need to have the effectiveness of precautions checked by measurement, the only certain way of finding what pollutants are present in the operator's breathing zone with a given combination of parent metal/consumables/operator/ventilation/work site
4. With processes not listed in Table 24.1, see the appropriate chapter of this book, and proceed as above with appropriate modifications
5. Where highly toxic materials such as cadmium or beryllium are involved, see Chapter 23 and elsewhere in this chapter for precautions to be taken. If the particular circumstance is not covered above, or if work will be prolonged, take expert advice on recommended procedures, precautions, and the need to check pollutant levels when starting work
6. If any worker shows ill-effects which may stem from fume, seek medical advice. Unless fume is ruled out as the cause, check pollutant levels

Table 24.1 Common oxygas and arc welding processes: current (1980) general recommendations for protection from fume

Processes (to which this Table is applicable)
Oxyacetylene
Manual metal-arc
Metal-arc gas-shielded
Tungsten-arc gas-shielded

Parent metal requiring minimum protection
Mild steel
Aluminium

Working conditions	Precautions
Open air	Stand upwind of weld. If fume is unpleasant or toxic use local extraction, dust respirator or other personal protection
Open shop	General ventilation. Also local extraction for all but occasional fork and for toxic materials. In addition, toxic material may demand personal protection for the welder and those nearby
Semi-confined space	Local extraction. Also personal protection for intensive work (high duty cycle or high current) or toxic materials, or if fume is unpleasant
Confined space	Local extraction and personal protection. Where the atmosphere presents an immediate hazard to life substantial protection is required, such as that offered by breathing apparatus and careful emergency planning. Check for fire and explosion hazards (Chapter 9) and the possibility of oxygen enrichment (Chapter 4)

General note: applicable to all working conditions. Be aware of any special hazards of parent material and consumables. If in doubt, obtain the recommendations of the manufacturer or supplier

Further information
A Welding Institute booklet, Ref.24.5, briefly indicates, for each of a range of processes, typical fumes, measured concentrations, specific pollutants, and precautions recommended in various situations. It will probably be revised as necessary to take account of new information on welding processes and revision of TLVs, so the latest available edition should be obtained. Reference 24.6 covers local extraction systems.

This information is based on an extensive research project at The Welding Institute, which is consequently able to offer a comprehensive service for the measurement of welding fume, to suggest measures for its reduction, if necessary, and to evaluate their effectiveness in practice. Some of the laboratories listed in Appendix 2, Part 2, will also be able to assist in a similar manner.

A wide range of protective equipment is marketed by welding manufacturers and distributors, by specialist ventilation manufacturers, and by safety equipment distributors.

See also the reference in Appendix 4 to the comprehensive report on welding fume produced by the Institute.

CHAPTER 25

LIGHTING

When work must be carried out in areas where insufficient daylight is available it will be necessary to provide lighting, which will almost invariably be electric. Two cases must be covered: normal operation and emergency lighting.

NORMAL LIGHTING
The IES Code, Ref.25.1, has the following entries in its General Schedule:

Welding and soldering shops

	Standard service illuminance, lux
Gas and arc welding, rough spot welding	300 (500)
Medium soldering, brazing, spot welding, e.g. domestic hardware	500 (750)
Fine soldering, spot welding, e.g. instruments	1000 (1500)
Very fine soldering, spot welding, e.g. electronics	1500 (3000)

It also gives further information on the application of these figures, with a note (page 61) to increase them (to the figures shown above in parentheses) if reflectances or contrasts are unusually low. Suitable types of lamp and glare factors permissible are also indicated.

The type of work encountered in the normal welding shop will come within the low reflectance clause, and so the figures in parentheses will be those applicable.

The Code also calls for the maintenance requirements to be assessed, with extra illumination being provided initially to allow for lamp deterioration and dust accumulation on luminaires (lighting fittings) during maintenance intervals between routine lamp replacement and cleaning. This proviso, too, will be particularly applicable to a normal welding shop.

Good general illumination will permit the use of somewhat lighter viewing filters, because the eyes adapt to general illumination (by the narrowing of the pupils), and the arc light has to be reduced less to match the general illumination. This gives the welder a better, more comfortable view of the weld, and renders him less susceptible to dazzle by an accidental view of an arc.

For work on site some welding generators are available with an outlet to power lights. As this is often of limited power, or of nonstandard voltage or frequency, the exact facilities required should be checked carefully against the specification. There is also a Lighting Guide for Building and Civil Engineering Sites, Ref.25.2.

For the illumination of fuel gas stores, where an explosive concentration could be caused by a leak, flameproof luminaires (and fittings such as switches) will be required. A possibility is to use a normal floodlighting luminaire outside the risk area, which may also have advantages from the point of view of security. Further information may be found in Refs 25.3 and 25.4.

EMERGENCY LIGHTING
If a complete electrical power supply failure occurs after dark, emergency lighting will be needed to ensure that workers are able to see well enough to carry out the following actions:

1   Make safe any radiographic equipment, especially isotope sources
2   Shut down all gas flames for welding, cutting, preheating, etc.
3   Switch off all electric welding equipment
4   Render safe any equipment relying on supplies also cut off by an electric power failure, such as water cooling, compressed air, or ventilation systems
5   Ensure that all crane motors are switched off and that any suspended loads presenting a hazard will be marked if necessary
6   Rescue anyone trapped, such as in a crane cab or lift
7   Evacuate the premises in an orderly fashion, making sure no one is left behind

Similar requirements apply if it is necessary to cut off the supply in the event of fire.

General guidance as to the design, installation, testing, and maintenance of a suitable system may be found in Ref.25.5.

CHAPTER 26

FIRE

It may be argued that fire safety is not the concern of welding personnel, but falls within the province of the safety officer and the fire brigade. But the risk of fire breaking out during welding is much greater than during other engineering workshop operations, so all concerned should do their best to minimise hazards to life and property. In addition, in the UK a number of statutory duties are placed on employers and employees by legislation which is steadily extending to cover all aspects of fire safety (see Chapter 28).

The general aims of fire safety procedures are, firstly, to prevent a fire breaking out, and, secondly, if it does break out, to:

Detect it
Rescue all persons at risk
Prevent it spreading
Extinguish it

FIRE PREVENTION
For a fire to start, three essentials must be present:

Combustible material
Oxygen
A source of ignition

In most instances of welding and cutting oxygen will be present in the form of air which the operator will breathe. The flame or arc and any hot particles emitted from welding or cutting operations act as an adequate source of ignition in most conditions, so if combustible material is nearby it may catch fire.

Combustible material
Flammable material which may be found near welding operations includes:

1   Parent metal and filler of magnesium or its alloys: swarf is especially

flammable, and may ignite spontaneously in the presence of oil; it is dangerous to use an unsuitable extinguisher
2   Wood: often found as blocks to support or wedge work
3   Glass reinforced plastic: though not normally considered flammable may be ignited rapidly by an oxyacetylene flame
4   Cardboard or other packing material
5   Oil or grease
6   Solvents: before changing to a nonflammable solvent the fume problems mentioned in Chapter 23 should be carefully considered
7   Dust: flammable dust may accumulate in ventilation systems, and once ignited the air flow will fan and distribute the flames; often a special hazard in site work
8   Insulation materials, especially foam plastic

Cigarette lighters

There is an unconfirmed report that workers have been seriously injured by fire and/or explosion originating from disposable cigarette lighters carried in their pockets while welding. There appear to be three ways in which the butane fuel could escape: firstly, by the plastic case cracking, secondly, by spatter penetrating the plastic case, thirdly, from a leaky valve. The third mechanism could well apply to a refillable lighter, with a leak developing after a period of use. Even matches could present a hazard if a box were ignited by a globule of spatter. Keeping matches or metal lighter in an inner pocket would offer a degree of protection, but for complete safety the ideal would be to avoid carrying these items on the shopfloor, a decision probably best left to the individual workers in the present state of knowledge.

Oxygen enrichment

In some welding and cutting operations, particularly gas cutting where ventilation is poor, the oxygen content of the air may be increased thus increasing the risk of fire. Oxygen enrichment and how to avoid it has been covered in detail in Chapter 4.

Smoking

As welding and cutting are such effective sources of ignition there is usually little point in banning smoking, with two exceptions. No smoking should be allowed:

When handling or connecting oxygen or fuel gas cylinders
When using solvents. Note that if the solvent is not flammable, it will probably be decomposed by heat to give toxic fumes (see Chapter 23).

Security

Finally, since many fires in the UK are started by intruders, good security is desirable to avoid fire as well as for its primary purpose.

DETECTION
Starting with the simplest and most reliable detector — the human being — it will be found that the welder will not generally be aware of a fire while arc welding, as it is not visible through his filter glass. To overcome this problem it is advisable to have a 'standby fireman' on duty when welding or cutting must be carried out in the vicinity of flammable material. If fire breaks out, he must interrupt the welder and be trained and equipped to take appropriate action.

Fire alarm systems, activating an alarm bell only or also sending a signal direct to the fire service, may be difficult to set to a suitable threshold of detection which will avoid false alarms, especially where preheating by flame or a furnace is in use: welding fumes may interrupt the light beam of a smoke detector.

Sprinkler systems combine detection and action: excess temperature triggers a sprinkler head to open, supplying a water spray to extinguish the fire. As these, too, are liable to false operation in a welding environment, care should be taken when installing them over or near machinery, drawing offices, inspection departments, etc. which contain items likely to be damaged by water. Where there is oil or metal powder which may catch fire, water will not extinguish these materials, and may spread the fire (see 'Extinguishing fires' below) so normal sprinklers must not be fitted. However, there are special high pressure sprinklers which can extinguish oil fires.

RESCUE
The first step is to warn all persons in the building that there is a fire. Ideally a fire alarm should be readily audible above the noisiest operation, and so distinctive that it cannot be mistaken for any other signal. In practice this may be impossible to achieve, and it will be necessary to warn individually those workers using particularly noisy processes.

Emergency exits may be required where there are not enough normal exits to ensure a speedy evacuation of a building wherever a fire may break out. A variety of catches are now available as standard items for exit doors with such facilities as opening by key or by breaking a glass tube, with built-in audible alarm or remote indication of opening, so that security can be fully maintained at unmanned emergency exits. The exits must be adequately signed, kept clear of rubbish, etc. to allow free exit even in dense smoke, and provided with emergency lighting if necessary.

A rescue plan will also need to take account of those who may have extra difficulty reaching the exit, for example disabled workers, visitors, or overhead crane drivers. It is essential to be able to check, by assembly and roll-call in most situations, that everyone is clear of the building, and to establish this as soon as possible.

Regular practice fire alerts will be helpful, both to check plans and avoid panic in an emergency.

PREVENTING THE SPREAD OF FIRE
The principle here is to confine the fire to as small a volume as possible. Fire check doors should be kept closed, particularly after working hours. If they prove too inconvenient, catches to hold them open are now available which will be released by a thermal trip in the event of fire.

Large continuous buildings are undesirable from the fire protection point of view, but are often essential for efficient industrial operations. Roof vents, which open automatically in the event of fire below them, are claimed to release smoke before it can spread and damage the rest of the building, and thus prevent it from obscuring the firefighter's view, allowing him to use the minimum effective amount of water.

Fire can spread readily through ventilation trunking from one space to another. To overcome this, cutoff flaps released by temperature rise may be fitted. In many installations it will be desirable to cut off the ventilation fans, but some buildings may be designed to use air pressure to blow smoke out of hazardous areas, and such systems must be left running.

In many instances of fire in the welding shop the electricity supply can be switched off at the incoming mains, cutting off all arcs and thereby alerting all workers, and avoiding the risk of electric shock to those fighting the fire. But it will be necessary to be sure that all relevant supplies to the area are cut off, and that doing so does not strand anyone in a lift, crane cab, etc., leave any equipment in a dangerous state, or cut off lighting needed to escape. Therefore, this should receive careful attention when planning actions in the event of fire.

EXTINGUISHING FIRES
As fires involving different materials require varying treatment for effective extinction, they have been classified (Ref.26.1):

Class A   Fires involving solid materials, usually of an organic nature, in which combustion takes place with the formation of glowing embers
Class B   Fires involving liquids or liquefiable solvents
Class C   Fires involving gases
Class D   Fires involving metals

Examples of the above classes are fires of the following materials found in the welding shop:

Class A   Wood, electrical insulation material, rags, dry paint film
Class B   Oil, grease, flammable solvents
Class C   Acetylene, propane
Class D   Magnesium, aluminium powder

Fixed fire extinguishing installations
These usually take the form of sprinkler systems, as discussed under 'Detection' above.

Alternatives are 'rising mains' in multistorey buildings, a pipe into which the fire brigade can pump water at ground level, drawing it off at whatever level needed; or hose reels, connected either to the normal water supply or to the rising main.

For the use of water, see below.

Portable fire extinguishers
Extinguishers in common use rely on various agents to provide the essential extinguishing actions of cooling and excluding oxygen. They can be classified into five types, depending on the agent:

Water
Carbon dioxide
Halon
Dry powder
Foam

As it is important to choose the right type of extinguisher to suit the fire, it is desirable to colour code them for easy identification. This is now covered by BS 5423 : 1981 (Ref.26.2). The colours normally used are: Water — red, Carbon dioxide — black, Halon — green, Foam — white, Dry Powder — blue. Powder suitable for Class D fires: silver.

Water extinguishers require some propulsive force, which may take the form of air compressed above the water (stored-pressure type), carbon dioxide gas in a sealed, replaceable cartridge (gas cartridge type), or the reaction of acid with bicarbonate solution (the obsolescent soda acid type). Water is suitable to extinguish Class A fires, but should not be used near live electrical equipment as the jet will conduct electricity to the operator, especially with the soda acid type. It is dangerous to use water on Class B and D fires: it may intensify or spread the fire, or cause explosions. Water may also be used from a bucket (Ref.26.3).

Carbon dioxide extinguishers are suitable for use on Class B fires, and on small Class A fires, particularly where equipment or work would be irretrievably damaged by water or powder. They may be used on electrical fires.

Halon extinguishers have a liquid active agent such as bromotrifluoromethane, with gas cartridge or stored pressure propulsion, as for water extinguishers. They are suitable for Class B fires but may be used for Class A fires, particularly where water or powder contamination is to be avoided. They may be used on electrical fires. The vapours are toxic to some extent and, like nonflammable solvents, tend to form more toxic compounds on heating, so care should be taken to avoid inhaling excessive quantities of vapour. If, for example, they have to be used in an enclosed space those exposed to the vapour should go out into the open air to recover as soon as possible.

Dry powder extinguishers again use a gas cartridge or stored pressure propellant. They are all suitable for use on Class B fires, some in addition being rated for Class A and/or D.

Foam extinguishers also have a gas cartridge or stored pressure propellant. They are suitable for use on Class B fires where the liquid is contained. They should be aimed not at the liquid but at the container wall above the liquid surface, allowing the foam to run gently down without agitating the liquid. They may not be used on live electrical equipment.

Extinguishers that are too large to be carried may be mounted on trolleys which can be wheeled to the fire, but hand sizes as well will be needed to get at less accessible places.

Class C fires, those involving flammable gases, are best extinguished by cutting off the gas supply. If the fire is extinguished by other means, the escaping gas may accumulate, then reignition (from a spark, hot surface, etc.) will cause a substantial explosion.

Guidelines for use of fire extinguishers
Do not start an attempt to extinguish a fire or do abandon any attempt already started if:

(a) the fire appears to be beyond control, or
(b) the escape route is threatened by fire, or
(c) the escape route is threatened by smoke

Choose the right extinguisher
Use it correctly

In any workplace it is important to know:

Location of emergency exits
Location of extinguishers
Types of extinguisher to be used on any likely fire
Operational procedure (push or pull knob, upright or inverted use, etc.)
Application (direct jet at or near base of fire etc.), if possible by practice

Ensure that extinguishers are available to place by any especially risky operation. All personnel should receive instruction in the use of all the available fire extinguishers: it is too late to read the instructions when the fire is well established.

If the fire appears to be extinguished, keep a close watch on the area until all possibility of reignition has been excluded.

Ensure that used extinguishers are promptly refilled or replaced.

Maintenance of fire extinguishers
Fire extinguishers must be subject to effective preventive maintenance, covering especially the avoidance of pressure loss in stored-pressure or gas-cartridge systems, or loss of gas in carbon dioxide extinguishers. The manufacturers of extinguishers usually provide such a service or, alternatively, operate a rental scheme.

Fire brigade
The fire brigade should always be notified of a fire immediately, however small it appears at first: in the UK, this is a legal requirement (Ref.26.4). The public and/or works fire brigade will also be able to advise on the following items on which they should be fully informed:

Fire precautions (in the UK a certificate is legally required)
Layout of works to reduce fire risks
Location of gas cylinder stores, bulk storage tanks, radioactive materials, and other special hazards
Notification of fire, e.g. access to telephone
Location of works
Site entrance and identification
Area location within large sites

The fire brigade should be kept informed of the works layout and hazard areas (such as flammable gas stores and radiographic isotope stores) by taking a representative round the plant after any major alteration or in any event at, say, two-year intervals.

Further information
A code of practice for fire extinguishing installations and equipment is being published as Ref.26.5.

CHAPTER 27

FIRST AID

INTRODUCTION

Although some large works will have a nurse or doctor on the staff to ensure prompt treatment of injuries, most factories will call in an ambulance or doctor in emergency, with immediate first aid treatment from a fellow employee. By far the best assurance of effective first aid (sustaining life, preventing the injury from becoming worse, and promoting recovery) is a properly trained and certificated 'first aider'. The UK law requires such first aiders, and appropriate equipment, in most premises subject to the Factories Act (see Chapter 28), and there are proposals to require these provisions more generally under the Health and Safety at Work etc. Act. Information on approved first aid treatment is given in Refs 27.1 and 27.2, but practice is essential for many aspects. This chapter therefore contains only brief notes on some of the problems particularly associated with welding.

First aid treatment for the following conditions will be described in the subsequent paragraphs:

Electric shock and burns
Eye injuries: foreign bodies
              burns
              chemicals
              arc eye
Major wounds
Minor cuts and abrasions
Burns of the skin
Traumatic shock
Heat stroke
Exposure to harmful gases and fumes

ELECTRIC SHOCK AND BURNS

The effects of electricity on the body are electric shock and/or electric burns.
Cause: the passage of electric current through the body.

Electric shock

When electric current passes through the body, the nature and severity of the injuries sustained by any individual depend upon the amount of current and its pathway through his body. This is in turn related both to the applied voltage and to the state of the body at the time. The risk to health is greatly increased if at the time of contact the individual is perspiring freely from his exertions, if his clothes and shoes are wet, or if, for instance, he is standing on a metal floor or on wet ground.

The victim of electric shock may be in a state of acute fear, or unconscious and in a condition of apparent death. Cessation of breathing or of the heart's action can occur from spasm and paralysis of the breathing muscles, and from the action of the electric current directly on the heart or on the control centres for the lungs and heart situated in the brain.

Treatment: switch off current immediately or send someone to do so. Do not attempt to remove a person from contact with high voltage unless suitable articles insulated for the system voltage are used for this purpose. When attempting to free a person from contact with low or medium voltage use rubber gloves, boots or mat, or insulated stick, but if these are not available use a loop of rope, cap, or coat to drag the person free. Sufficient insulation for normal arc welding voltages will be given by dry clothes.

Whatever is used should be dry and nonconducting. *Apply artificial respiration immediately,* making sure that the injured person's airway is clear: the most effective is mouth-to-mouth (mouth-to-nose). Use external heart compression if the pulse has stopped.

All employees responsible for the administration of first aid must be familiar with modern methods of resuscitation. Since speed is vital, it is advisable for those concerned with electric arc welding to be trained either with a full first aid course, or in resuscitation only. Even if the victim seems to recover rapidly, he should be seen by a doctor as soon as possible.

Electric burns

These burns usually occur at the point of entry of the electric current and at its exit. Their area is small but they tend to be deep and painful and may severely damage the affected tissues.

Treatment: they should be treated in the same way as burns and must always be seen by a doctor or nurse.

EYE INJURIES

Foreign bodies in the eye
Cause: dust and metallic particles.

Assessment: to examine the eye carry out the following procedures:
(a) seat the patient in a good light
(b) stand behind the patient and support his head against your body; have the patient's head tipped well back
(c) hold the eye open with two fingers, one on each lid, and ask the patient to look in all directions while you examine the eye. The use of a magnifying glass is helpful
(d) pull the lower lid down to inspect its inner surface; the upper lid can be everted providing the first aid attendant has been instructed in procedure

Normally, foreign bodies lodge on the central surface of the eye, on the white of the eye or on the inside of a lid and, in the great majority of cases, are easily removed. If complaints of pain continue and no particles are visible, the case should be referred to a nurse or doctor. A metallic particle may also penetrate into the interior of the eye, leaving little evidence on the surface of its penetration, except to a qualified medical practitioner or nurse. Penetration of the eye is a serious emergency and pain in the eye, disturbance of vision, and distortion of the pupillary shape should be regarded as evidence that this has occurred.

Treatment: particles on the surface of the eye:

(a) wash the eye, using a sterile saline solution in a disposable container, or an eye bath
(b) if a particle still remains on the surface, make a pointed spill of moistened cotton wool, or use a moist readymade pledget, and sweep the surface of this over the eye. If the particle does not come away it has partially penetrated the surface, and the injured person should be referred to a doctor or nurse. Before doing so, cover the eye with an eyepad
(c) *do not use the corner of a handkerchief, unless laundered clean, to remove a particle of dust; otherwise you introduce infection into the eye*

Burns of the eye
Cause: from the spatter of hot particles.

Assessment: a small grey opaque area may be seen on the surface of the eye and the patient will often complain of soreness and pain of the eye.

Treatment: introduce Castor Oil BP into the eye sac, using a 60 microlitre

(1 minim) single dose capsule. Cover the eye with a light eyepad, and refer the patient to a doctor or nurse.

Chemicals in the eye
Cause:
1   Splashes of acids, alkalis, solvents, and other chemicals
2   Introduction into the eye of chemical dusts such as lime, cement, and calcium carbide
3   Exposure to irritant gases, fumes, and vapours

Assessment: irritation and pain of the eye, redness and watering, hypersensitivity to light, clouding of the clear front surface of the eye.

Treatment: immediate efficient first aid treatment can be the means of saving the injured person's eyesight. Copiously flood the eye with cold water, using the nearest source, for at least ten minutes: disposable containers holding sterile saline solution can be purchased and should be used only once. If the patient has difficulty in opening his eye because of eyelid spasm, the eyelids should be held apart. Introduce Castor Oil BP into the eye sac, using a 60 microlitre (1 minim) single dose capsule. Obtain the advice of a doctor or nurse, or send the injured person to hospital.

Arc eye
Cause: exposure of the unprotected eye to ultraviolet radiation.

Assessment: symptoms usually develop 4-8 hours after exposure and consist of soreness of the eye, a burning sensation, and watering and hypersensitivity to light. The eye becomes inflamed and the patient often passes a disturbed night. The condition does not usually lead to any serious permanent aftereffects.

Treatment: apply cold compresses to the eyes and irrigate them with an astringent lotion, then give the patient dark glasses to wear. The astringent lotion, which can be obtained from a local dispensing chemist, has the following formula:

Acid boric 650mg
Zinc sulphate 32mg
Liq. adrenaline hydrochloride 1/1000 0.6mlitre
Distilled water 31g

The lotion can also be used at home. Do not use any eye drops: there are special eye drops for this condition but they must be prescribed by a doctor or nurse.

Enquire from your supplier or local chemist concerning the frequency with which arc eye lotion and saline solution should be changed.

Contact lenses, if worn, should be removed when symptoms become apparent, or if gross overexposure has occurred, and not worn again until recovery is complete.

MAJOR WOUNDS
Severe bleeding from major wounds must be stopped immediately. Apply pressure with the fingers, over a dressing if immediately available. If the wound is large or gaping, press the edges together; maintain the pressure for 5-15 minutes. Send for a doctor or nurse, or get the injured person to hospital.

MINOR CUTS AND ABRASIONS
The importance of good first aid treatment is the prevention of infection, for an infected wound can incapacitate the injured person for many weeks and possibly result in a permanent disability.

Treatment: wash the wound with water, or with soap and water if the dirt is present around the wound, or clean with chlorhexidine gluconate in the appropriate dilution; the latter has an antiseptic action but does not injure the tissues. Dry the surrounding skin with fresh pieces of cotton wool, apply a clean sterile prepacked dressing to the wound, and a bandage or adhesive plaster. If a wound is gaping, refer the injured person to a doctor or nurse because it may require stitching.

Do not use such antiseptic solutions as iodine, or acriflavine, because they damage the tissues and can produce skin rashes.

BURNS OF THE SKIN
Cause: dry or wet heat, or chemicals, particularly acids and alkalis.

Assessment: the seriousness of any burn depends on the following factors: area, depth, part of body affected, and age and state of health of the patient. The more severe the burn the more likely it is that the patient will be affected by shock, which stems mainly from loss of fluid from the raw surface of the burn and into the damaged tissues. The burn will show itself as red skin with or without blisters, or various layers of the skin and underlying tissues will have been removed.

Treatment: the objectives of treatment are: to prevent shock, avoid infection, and relieve pain. The first aider should not continue treatment of any burn that is larger than 12mm in diameter but refer the patient to a doctor or nurse for further advice.

Dry or wet burns 12mm diameter or less
Flush and clean the burn with cold water for at least ten minutes until the pain

lessens, and then cover with a clean sterile prepacked dressing. The sterilised dressing can be left in position for two to four days, and only the dirty outside covering needs to be replaced. If hot plastic material becomes attached to the skin it should not be disturbed by the first aider. Refer to a nurse or doctor at once.

Do not prick any blister or remove its overlying skin.

Dry or wet burns over 12mm diameter
Flush the burn with cold water as above. Cover the burn or burns with clean or sterilised dressings or with clean towels and get the patient to a doctor or hospital as quickly as possible.

Chemical burns
Speed of action is of the utmost importance. Remove any contaminated clothing and flood the affected areas with water from the nearest available source for at least ten minutes. Cover the injured areas with sterile dressings or clean laundered towels and transport the patient immediately to hospital.

TRAUMATIC SHOCK
True shock, which should be distinguished from a faint, is caused by loss of body fluid: by loss of blood or serum from the outer surface of the body or from the internal organs and from vomiting and excessive sweating.

The shocked patient will be anxious, pale, cold, clammy, and restless, his breathing will be shallow, his pulse will be rapid and he will complain of thirst.

Treatment: the only treatment is replacement of lost fluid, and this has to be carried out in hospital. The first aider must prevent the shock getting worse and expedite the patient's transfer to hospital. The preventive measures may be summarised as:

1. Call an ambulance
2. *Move the patient as little as possible and move as gently as possible.* The best position for the patient is the semi-prone position with the head at a lower level than the legs
3. Check any external haemorrhage
4. Loosen tight clothing
5. Keep the patient warm but do not overheat. *Do not use hot water bottles*
6. *Do not give the patient fluids to drink:* if he is thirsty he can rinse his mouth with water and spit it out
7. Be cheerful and confident and make no statements concerning the patient's condition in his hearing

A person who has fainted appears to be in a state of shock, but recovery and normal appearance occur within a few minutes.

HEAT STROKE
Treatment: cool the patient by applying cloths or other absorbent material soaked in cold water. Send him to hospital.

EXPOSURE TO HARMFUL GASES AND FUMES
If adequate precautions are not taken the welder may be exposed to harmful concentrations of certain gases and fumes, particularly:

Carbon monoxide
Oxides of nitrogen
Ozone
Phosgene
Metal fumes

These, their sources, and some of their toxic effects, are described in Chapter 23.

Carbon monoxide
The effect of this gas is to diminish the oxygen-carrying capacity of the blood. It renders the individual dizzy and weak, complaining of headache and impaired mental concentration, and it may lead to a state of unconsciousness.

Treatment: remove the victim from the dangerous atmosphere to the fresh air, loosen tight clothing around the waist and neck, remove any dentures, give artificial respiration, and, if available, administer a mixture of 95% oxygen and 5% carbon dioxide. Send for a doctor and an ambulance.

Oxides of nitrogen, ozone, phosgene
These gases are being considered together because, broadly speaking, they produce similar harmful effects. Unfortunately they give little warning that they have been inhaled in excessive amounts, other than irritation of the nose and throat, but 24-28 hours later they may produce a severe inflammation of the lungs, so that the lung tissues become overloaded with water and the affected individual undergoes a state of asphyxiation.

Treatment: if excessive exposure is suspected the patient should be kept at absolute rest, lying down, until such time as he has been seen by a doctor. Oxygen should be administered at a flow rate of 3-10 litre/min, *but artificial respiration must not be given.*

Metal fumes
A short but acute illness (metal fume fever) occurs when fume is inhaled from a metal heated to a temperature above its melting point. Many metals can cause this condition but zinc, copper, and magnesium are the most likely to do so. Some hours after inhaling the fume the individual complains of tiredness, headache, aching pains in his muscles, thirst, a sore throat, a cough, and sometimes of a tight feeling of his chest. He will develop a high temperature, have shivering attacks, and perspire profusely. Complete recovery will normally occur in 24-48 hours, and there are not usually any permanent injurious aftereffects.

If a worker suffers from metal fume fever, the environmental cause should be investigated and remedied. If the pollutant is not positively identified, medical advice should be sought immediately.

Treatment: the sick person should go to bed immediately, keep warm with blankets and a hot water bottle, have no solid food but drink plenty of fluids. If symptoms continue for more than 24-48 hours, he should be seen by a doctor.

Cadmium
The inhalation of cadmium fumes gives rise to symptoms akin to metal fume fever, in some cases followed by acute inflammation of the lungs. Immediate medical advice should be sought.

General
Any persistent difficulty in breathing should be investigated by a nurse or doctor.

FIRST AID EQUIPMENT
All first aid equipment should be kept in a suitable cabinet in a clean and easily accessible location. *Dirty dressings are a source of infection.*

Stretchers should be provided where necessary, and provision made for urgent transport to hospital.

First aid equipment should be regularly checked to confirm that it is present and appears in good order, that stretchers are sound, and that any instructions, such as emergency telephone numbers, are up to date.

CHAPTER 28

LEGISLATION ON WELDING SAFETY IN THE UK

INTRODUCTION
The UK legislation affecting welding safety will be briefly reviewed in this chapter, starting with general Acts (covered only as far as they relate to welding, and not in their overall provisions) and proceeding to specific regulations. Finally, the implications of the Employment Protection Act, and of civil actions for damages, are discussed.

Before the law can provide for safety the sources of danger must be identified; suitable protective measures must then be established, or the decision made to ban the process concerned.

THE HEALTH AND SAFETY AT WORK ETC. ACT, 1974
This is an 'enabling' Act; it states duties in general terms, and enables specific regulations to be made.

Employers and self-employed persons must ensure, as far as is reasonably practicable, the health, safety, and welfare at work of all employees and the health and safety of other persons. In particular, they must:

(a) provide and maintain safe plant and systems of work
(b) arrange safe use, handling, storage, and transport of articles and substances
(c) provide the necessary information, instruction, training, and supervision
(d) maintain a safe place of work, and provide and maintain safe means of access
(e) provide and maintain a safe working environment

The designer, manufacturer, importer, or supplier of any article or substance for use at work is required to ensure that it is safe if properly used, that it is tested as necessary, and that adequate information is provided.

Employees must take reasonable care for their safety and that of others, and cooperate with employers.

Enforcement of the Act is generally through Inspectors appointed by the Health and Safety Executive, who are given wide-ranging powers of investigation and powers to issue:

(a) improvement notices, requiring the improvement of conditions, which in the Inspector's opinion contravene a specific regulation, within a stated period
(b) prohibition notices, prohibiting activities which in the Inspector's opinion involve a risk of serious personal injury, either with immediate effect or after a stated period

The law provides for an appeal to an industrial tribunal, to ensure that the final decision does not rest with Inspectors; an improvement notice is suspended while an appeal is proceeding but a prohibition notice is not. However, it will be far more productive to remedy an unsafe situation after amicable discussion with the Inspector than to be forced to take action by a notice: his aim will be to achieve safe working conditions by reasoned argument.

A problem which is sometimes encountered is a variation in the interpretation of the law by Inspectors in different parts of the country. This may be reduced as more requirements are spelt out in detailed regulations, rather than being left to the Inspector's discretion.

The Health and Safety Executive issues advice in the form of leaflets, booklets, etc. (of which those relevant to welding and allied processes are listed in Appendix 1).

As a last resort the Inspector may bring a prosecution in the Courts, which may impose fines or terms of imprisonment on those found guilty.

The Act does not repeal earlier legislation such as the Factories Act, 1961, but aims to replace it progressively by new regulations and codes of practice. In respect of fire safety it provides that new Fire Certificates shall be issued under the Fire Precautions Act, 1971. Enforcement of earlier Acts is now the responsibility of the Inspectors of the Health and Safety Executive.

THE FIRE PRECAUTIONS ACT, 1971
This requires the provision of precautions against fire, and is implemented by requiring the premises it covers to be inspected, usually by the local Fire Authority, and a certificate to be obtained.

THE FACTORIES ACT, 1961
This Act consolidates earlier Factories Acts. Its provisions apply only to factories

(the purpose of the Health and Safety at Work etc. Act above is to make further provisions to secure the health and safety of persons in all places of work).

Firstly, it has specific requirements. Section 4 requires adequate ventilation of each workroom to render harmless all fumes from work. Section 30 requires testing and certification by a responsible person before entry into any confined space in which dangerous fumes are likely to be present. Section 31 requires precautions against explosive or inflammable dust, gas, vapour, or substance, specifically in subsection 4: no plant, tank, or vessel which contains or has contained any explosive or flammable substance shall be welded, brazed, soldered, or thermally cut until all practicable steps have been taken to remove the substance or render it harmless. Section 63 requires exhaust appliances, and all other measures where practicable, to prevent injurious or offensive fumes entering the air of a workroom, and to prevent their accumulation or inhalation.

Secondly, it empowers the making of specific regulations, of which those with particular reference to welding are discussed below. Though the letter of the law applies only to premises subject to the Factories Act, the various regulations may be taken as a yardstick for safe working conditions to be achieved elsewhere under the Health and Safety at Work etc. Act.

THE ELECTRICITY REGULATIONS, 1908 (Statutory Rules and Orders No. 1312, amended by No. 739 of 1944)

These regulations, made under earlier Acts, continue in force under the Factories Act, 1961, and require all electrical equipment to be constructed, installed, protected, worked, and maintained to prevent danger. Exemption 1 exempts systems in which the voltage does not exceed 250V DC or 125V AC from most insulation requirements. Exemption 4 exempts apparatus used for electrothermal processes, which covers most forms of arc welding, provided the necessary precautions are taken to prevent danger.

THE BLASTING (CASTINGS AND OTHER ARTICLES) SPECIAL REGULATIONS, 1949 (Statutory Instrument No.2225)

These regulations, made under earlier Acts, continue in force under the Factories Act, 1961. They prohibit the use of sand or other substance containing free silica as an abrasive. The remainder of the regulations apply only to blasting in the cleaning of castings, though they may be taken as guidelines where other articles are blasted. Work must be done in a special ventilated enclosure. Employees working in the enclosure must be provided with, and use, gauntlets, overalls, and helmets with a fresh air supply of not less than $0.17 m^3/min$ ($6 ft^3/min$).

THE FIRST AID BOXES IN FACTORIES ORDER, 1959 (Statutory Instrument No. 906)

THE FIRST AID (Standard of Training) ORDER, 1960 (Statutory Instrument No. 1612)
Now repealed by the Health and Safety (First Aid) Regulations 1981 (see Appendix 4)

THE SHIPBUILDING AND SHIP-REPAIRING REGULATIONS, 1960 (Statutory Instrument No. 1932)
Part V of the regulations, made under the Factories Act, 1961, covers precautions against asphyxiation, injurious fumes, or explosions in confined spaces. It deals with these under the headings of ventilation, flammable gas or vapour, oxygen deficiency, dangerous fumes (requiring, where there is reason to anticipate a risk, the provision of breathing apparatus, safety lamps, and rescue harness, and the training of employed persons in its use and in artificial respiration), gas cylinders, and acetylene generators. They continue with specific reference to cutting and welding in regard to equipment and precautions after use (specifically the disconnection and removal of gas cylinders and hoses when work ceases for the day or for a substantial period, except a meal interval).

THE CONSTRUCTION (GENERAL PROVISIONS) REGULATIONS, 1961 (Statutory Instrument No.1580, amended by No.94 of 1966, and No.1681 of 1974)
These regulations were originally made under earlier Acts, but continue under the Factories Act, 1961; they apply to building and construction works. Regulation 20 requires the adequate ventilation and testing of every excavation or enclosed space to avoid poisoning or asphyxiation. Regulation 22 requires stationary internal combustion engines used in an enclosed or confined space to be exhausted into the open air unless the ventilation is adequate to prevent danger to health.

THE IONISING RADIATIONS (SEALED SOURCES) REGULATIONS, 1969 (Statutory Instrument No.808, amended by No.36 of 1973)
These regulations made under the Factories Act, 1961, apply in premises subject to that Act. They will apply both to radiography and to electron-beam welding. They require the employer to appoint a 'competent person' to supervise operations, the keeping of registers and records of 'classified persons' exposed to radiation, and the notification to the Factory Inspectorate of the acquisition of a source and of any accidents. Classified persons will normally be required to wear dosemeters.

THE ABRASIVE WHEELS REGULATIONS, 1970 (Statutory Instrument No.535)
These, made under the Factories Act, 1961, require the maximum permissible speeds of abrasive wheels to be specified, machines to be marked with their

operating speeds, and guards and rests to be fitted, where practicable. Wheels must be fitted by competent persons who have been trained: competent persons must be appointed by the occupier of the factory and listed in a register which he must keep. The regulations will apply to portable grinders used on workpieces for the preparation and finishing of welds, and for fixed machines used to shape tungsten electrodes for TIG-welding.

THE HIGHLY FLAMMABLE LIQUIDS AND LIQUEFIED PETROLEUM GASES REGULATIONS, 1972 (Statutory Instrument No.907)

These, made under the Factories Act, 1961, will apply to propane used as a fuel gas, and to inflammable solvents with a flashpoint below 32°C. They require safe storage, marking of storage accommodation and vessels, and appropriate fire precautions.

THE PROTECTION OF EYES REGULATIONS, 1974 (Statutory Instrument No.1691, amended by No.303 of 1975)

These were made under the Factories Act, 1961, and require approved eye protection to be provided and maintained for all employed persons engaged in most of the welding and allied processes described in this book. Eye protectors must be issued permanently if the process is used more than fifteen minutes a day, or on more than two days a week. Employees must use the eye protection, take care of it, and report loss or damage.

Certificate of Approval No.1 (21 March 1975) issued under the regulations approves general protectors to BS 2092, and approves for welding and cutting processes protectors to BS 1542 and BS 679.

THE EXPLOSIVES ACT, 1875

This Act regulates the manufacture, keeping, sale, carriage, and importation of gunpowder and other explosive substances. It is still in force and has been confirmed and revised in minor details by the Health and Safety at Work Act. Note that it is not restricted to application to factories or workplaces: it will apply, for example, even where welding is being done privately, such as the repair of the welder's own car.

ORDERS IN COUNCIL (No.30) PROHIBITING THE MANUFACTURE, IMPORTATION, KEEPING, CONVEYANCE, OR SALE OF ACETYLENE WHEN AN EXPLOSIVE AS DEFINED BY THE ORDER, 1937 (Statutory Rules and Orders No.54, amended by No.805 of 1947)

This order defines acetylene, when mixed with air or oxygen (except in a torch), or at a pressure over $9lb/in^2$ (0.62 bar), as an explosive for the purpose of the Explosives Act, 1875, requiring a licence. A general exemption has been made for acetylene in approved cyclinders. Recently it has come to notice that many acetylene regulators, currently supplied and used, can be set to deliver gas at over 9lb/in² (0.62 bar), and the supplier, with user employers and employees, would be liable to prosecution if gas were to be used above this pressure without an HSE licence. It has been suggested that employees should be warned, and warning labels attached to the regulators, with future production of regulators modified to limit the available outlet pressure.

THE PETROLEUM (CONSOLIDATION) ACT, 1928
This sets up a licensing and control system for petroleum spirit, extendable to other substances.

THE PETROLEUM (CARBIDE OF CALCIUM) ORDER, 1929 (Statutory Rules and Orders No.992, amended by No.1442 of 1947)
This order applies the Petroleum (Consolidation) Act, 1928, to carbide of calcium restricting its keeping and use:

(a) 5lb (2.3kg) may be kept in separate hermetically closed metal vessels containing not more than 1lb (0.45kg) each
(b) 28lb (12.7kg) may be kept in an hermetically closed metal vessel or vessels in a dry place with unauthorised access prevented; notice must be given to the Local Authority. A fixed generator must have the maker's instructions posted by it

THE PETROLEUM (COMPRESSED GASES) ORDER, 1930, No.34
This order applies the above Act to named compressed gases including air, argon, hydrogen, oxygen, and nitrogen.

THE GAS CYLINDER (CONVEYANCE) REGULATIONS, 1931 (Statutory Rules and Orders No.679 amended by No.1594 of 1947 and Statutory Instrument No.1919 of 1959)
These regulations require gas conveyed by road to be in approved cylinders, colour-coded as below, and at a maximum pressure of 3000lb/in² (207 bar):

Air — grey
Argon — blue
Hydrogen — red
Nitrogen — dark grey with black band
Oxygen — black

EMPLOYMENT PROTECTION ACT, 1974
This gives protection to employees against unfair dismissal. Dismissal may be held to be fair if it has been occasioned by unsafe working of the employee. Two points should be borne in mind. Firstly, the employer should state clearly the safety rules, and the consequences of disobeying them, for example: 'No

smoking in the acetylene cylinder store; offenders are liable to severe disciplinary action, including instant dismissal without prior warning'. Such statements should appear in a clear notice and/or be included in a rule book issued to, and signed for by, each employee. Secondly, breaches of rules should be dealt with immediately, fairly, and consistently.

Under the Health and Safety at Work etc. Act, 1974, HSE Inspectors have the power to prosecute employees who endanger themselves or fellow workers. This unusual step would need to be supported by appropriate evidence that the defendant had had clear written warnings about his activities.

CIVIL PROCEEDINGS
As well as the above legislation, which carries the usual sanctions of fine or imprisonment for infringement, a further inducement to safe methods of working may be provided by the risk of having to pay damages in a civil action. Such an action is reported in Ref.28.1.

Further information
Copies of the relevant Acts and Regulations mentioned above can be purchased from Her Majesty's Stationery Office, or consulted in some major libraries. They are not listed separately in Appendix 1, but Refs 28.2 and 28.3 contain a general review of the legislation. References 28.4-28.8 give clear guides to particular Acts and Regulations. Reference 28.9 provides guidelines for the effective organisation of safety inspections in conjunction with employee representatives under the Health and Safety at Work Act.

APPENDIX 1
BIBLIOGRAPHY OF FURTHER INFORMATION

INTRODUCTION

This Appendix lists printed and other information supplementing that in the main text of the book. The aim has been to choose material which is helpful, practical, and readily available where possible, rather than research papers. To this end, the Appendix is divided into 5 main sections as follows:

1. The initial reference corresponds to the chapter
2. A list of relevant titles by series follows
3. The Welding Institute publications etc. are listed
4. Other teaching and instructional material is listed
5. The Appendix concludes with short notes on how to obtain access to this material

Various organisations from which advice may be obtained are listed in Appendix 2. To save space a number of abbreviations have been used:

BCGA   British Compressed Gases Association
BS   British Standard
HMSO   Her Majesty's Stationery Office
HSE   Health and Safety Executive (UK Government organisation)
TWI   The Welding Institute

REFERENCES BY CHAPTER

Chapter 1 — Process fundamentals

1.1   HOULDCROFT, P.T. 'Welding process technology'. London, Cambbridge University Press, 1977.

1.2   BS 499 : Pt 1 : 1965 'Welding, brazing, and thermal cutting glossary'. London, British Standards Institution.

1.3   DAVIES, A.C. 'The science and practice of welding', 7th ed. London, Cambridge University Press, 1977.

Chapter 2 — Gas cylinders and gas welding and cutting equipment

2.1 BS 349 : 1973 'Identification of contents of industrial gas containers'. London, British Standards Institution.
(Cylinder stores: positive recommendations are found in Ref.2.2 which give a useful guide, even though this pressure will not be reached in practice.)

2.2 HSE. 'Code of practice for the use of acetylene at pressures between 0.6 and 1.5 bar' (may become regulations in due course).
(Propane: Refs 2.3-2.5 apply.)

2.3 The Highly Flammable Liquids and Liquefied Petroleum Gases Regulations, SI 1972 no.917. London, HMSO, 1972.

2.4 LPGITA. 'Installation and maintenance of bulk LPG storage at consumers' premises. Code of practice 1'. Chertsey, Liquefied Petroleum Gas Industry Technical Association, 1961. (Pt 7 'Keeping LPG in cylinders'; Pt 10 'Fire precautions'.)

2.5 HSE. 'The storage of liquefied petroleum gas at factories', Health and Safety at Work Booklet 30. London, HMSO.

2.6 BCGA. 'The safe working, maintenance, and repair of gas cylinder and pipeline regulators used with compressed gases for welding, cutting, and related processes', CP1. London, British Compressed Gases Association, 1975.

2.7 BS 5741 : 1979 'Specification for pressure regulators used in welding, cutting, and related processes'. London, British Standards Institution.

2.8 BS 1780 : 1960 'Bourdon tube pressure and vacuum gauges. Appendix D — Precautions relating to gauges for use with high pressure gas; Appendix E — Precautions relating to gauges for use with oxygen and acetylene'. London, British Standards Institution.

2.9 BCGA. 'The safe working, maintenance, and repair of hand-held oxygen and fuel gas blowpipes used for welding, cutting, and related processes', CP2. London, British Compressed Gases Association, 1978.

2.10 BS 5120 : 1975 'Rubber hose for gas welding and allied processes'. London, British Standards Institution.

2.11 BS 1845 : 1977 'Specification for filler metals for brazing'. London, British Standards Institution.

2.12 BS 1389 : 1960 'Dimensions of hose connexions for welding and cutting equipment'. London, British Standards Institution.

2.13 HSE. 'Welding and flame cutting using compressed gases', Health and Safety at Work Booklet 50. London, HMSO, 1978.

Chapter 3 — Low pressure plant

3.1 BAA. 'The storage of calcium carbide', 3rd rev. ed. London, British Acetylene Association (now British Compressed Gases Association), 1962.

3.2 The Explosive Acts, 1875 and 1923.

3.3 SR and O no.992 The Petroleum (Carbide of Calcium) Order, 1929, amended by no.1442, 1947.

Chapter 4 — Safety precautions during gas welding and cutting operations
4.1 HSE. 'Fires and explosions due to the misuse of oxygen'. London, Health and Safety Executive.
4.2 HSE. 'Welding and flame cutting using compressed gases', Health and Safety at Work Booklet 50. London, HMSO, 1978.

Chapter 5 — Protective clothing, and eye and head protection, for gas welding and cutting
5.1 BS 2653 : 1955 'Protective clothing for welders'. London, British Standards Institution.
5.2 BS 1651 : 1966 'Industrial gloves'. London, British Standards Institution.
5.3 BS 679 : 1959 'Filters for use during welding and similar industrial operations (viewing)'. London, British Standards Institution.

Chapter 6 — The care of arc welding and cutting equipment
6.1 BS 638 : Pt 4 : 1979 'Specification for welding cables.' London, British Standards Institution.
6.2 BS 638 : 1966 'Arc welding equipment, Section 11: Plugs and sockets'. London, British Standards Institution.
6.3 BS 638 : 1966 'Arc welding equipment, Section 8: Electrode holders'. London, British Standards Institution.

Chapter 7 — Safety precautions during arc welding and cutting operations
7.1 BALCHIN, N.C. 'Electrical safety in welding'. *Welding Inst. Research Bull.*, **18** (3), 1977, 73-8; (4), 1977, 99-104; (9), 1977, 240.
7.2 LYON, T.L. et al. 'Evaluation of the potential hazards from actinic ultraviolet radiation generated by electric welding and cutting arcs'. US Army Environmental Hygiene Agency, AD/A-033 768 (available from Microinfo Limited, PO Box 3, Newman Lane, Alton, Hants GU34 2PG).

Chapter 8 — Protective clothing, and eye and head protection, for arc welding
8.1 BS 2653 : 1955 'Protective clothing for welders'. London, British Standards Institution.
8.2 BS 1651 : 1966 'Industrial gloves'. London, British Standards Institution.
8.3 BS 1542 : 1960 'Equipment for eye, face, and neck protection against radiation arising during welding and similar operations'. London, British Standards Institution.

8.4 SLINEY, D.H. and WOLBARSHT, M. 'Safety with lasers and other optical sources'. New York, Plenum Press, 1980.
8.5 BS 679 : 1959 'Filters for use during welding and similar industrial operations (viewing)'. London, British Standards Institution.
8.6 LYON, T.L. et al. 'Evaluation of the potential hazards from actinic ultraviolet radiation generated by electric welding and cutting arcs'. US Army Environmental Hygiene Agency, AD/A-033 768 (available from Microinfo Limited, PO Box 3, Newman Lane, Alton, Hants GU34 2PG).
8.7 BS 5330 : 1976 'Method of test for estimating the risk of hearing handicap due to noise exposure'. London, British Standards Institution.
8.8 DEPARTMENT OF EMPLOYMENT. 'Code of practice for reducing the exposure of employed persons to noise'. London, HMSO, 1972.

Chapter 9 — Precautions for welding and cutting vessels which have held combustibles

9.1 'Sheffield explosion'. London, *The Times*, 25 October 1973.
9.2 DEPARTMENT OF EMPLOYMENT 'Repair of drums and small tanks: explosion and fire risk', Health and Safety at Work Booklet 32. London, HMSO, 1975.
9.3 KLETZ, T. 'Safety in chemical operations' (contribution in). Chichester, John Wiley and Sons Limited.
9.4 DEPARTMENT OF EMPLOYMENT 'The safe cleaning, repair, and demolition of large tanks for storing flammable liquids', Technical Data Note 18. London, HMSO, 1974.
9.5 DEPARTMENT OF EMPLOYMENT. 'Entry into confined spaces: hazards and precautions', Technical Data Note 47. London, HMSO, 1974.
9.6 HSE. 'Hot work: welding and cutting on plant containing flammable materials', Pamphlet HS(G) Series no.5. London, HMSO, 1979.

Chapter 10 — Plasma arc processes
10.1 BS 679 : 1959 'Filters for use during welding and similar industrial operations (viewing)' London, British Standards Institution.

Chapter 11 -- Electro-slag welding and consumable-guide welding
11.1 BS 679 : 1959 'Filters for use during welding and similar industrial operations (viewing)'. London, British Standards Institution.

Chapter 13 — The thermit process
13.1 BS 679 : 1959 'Filters for use during welding and similar industrial operations (viewing)'. London, British Standards Institution.

Chapter 14 — Electron-beam welding
14.1 'Limit switches and interlocks' in 'Electrical accidents and their causes', 30-47. London, HMSO, 1958.
14.2 'Failure to safety of electrical control gear' in 'Electrical accidents and their causes', 26-52. London, HMSO, 1961.
14.3 The Ionising Radiations (Sealed Sources) Regulations, SI 1969 no.808. London, HMSO, 1969.
14.4 ANON. 'EB welding machine to Italy'. *Metal Constr.*, **11** (4), 1979, 169.

Chapter 15 — Laser welding and cutting
15.1 Protection of Eyes Regulations, SI 1974 no.1681. London, HMSO, 1974.
15.2 BS 4803 : 1972 'Guide on protection of personnel against hazards from laser radiation'. London, British Standards Institution.
15.3 HARRY, J.E. 'Industrial lasers and their applications. Chapter 6: Safety'. London, McGraw-Hill, 1974.
15.4 ANSI Z136.1-1973 'Standard for the safe use of lasers'. New York, American National Standards Institute, 1973.
15.5 LIA. 'Laser safety guide'. Cincinnati, Laser Institute of America, 1977.
15.6 SLINEY, D.H. and WOLBARSHT, M. 'Safety with lasers and other optical sources'. New York, Plenum Press, 1980.

Chapter 16 — Brazing and braze welding
16.1 'Precautions in the use of nitrate salt baths', Form 848. London, HMSO.
16.2 'Cautionary notice: nitrate salt baths', Form 849. London, HMSO.
16.3 The Highly Flammable Liquids and Liquefied Petroleum Gases Regulations, SI 1972 no.917. London, HMSO, 1972.

Chapter 18 — Thermal spraying
18.1 The Blasting (Castings and Other Articles) Special Regulations, 1949.
18.2 BS 5345 'Code of practice for the selection, installation, and maintenance of electrical apparatus for use in potentially explosive atmospheres (other than mining applications or explosive processing and manufacture)'. London, British Standards Institution.
18.3 BS 5490 : 1977 'Specification for degrees of protection provided by enclosures (electrical equipment)'. London, British Standards Institution.

Chapter 19 — Welding and flame spraying plastics
19.1 BEAMA. 'Safe usage of induction and dielectric heating equipment', Publication no.188. London, British Electrical Appliance Manufacturers Association, 1962.

Chapter 20 — Radiographic inspection

20.1 The Ionising Radiations (Sealed Sources) Regulations, SI 1969 no.808. London, HMSO, 1969.

20.2 DEPARTMENT OF EMPLOYMENT, HM FACTORY INSPECTORATE. 'Code of practice for site radiography'. London, Kluwer-Harrap, 1975. (This Reference is of more general application than the limited field indicated by its title.)

20.3 BS 3510 : 1968 'A basic symbol to denote the actual or potential presence of ionising radiation'. London, British Standards Institution.

Chapter 21 — Mechanical hazards

21.1 HSE. 'Safety in pressure testing', General Series GS/4. London, HMSO, 1976.

21.2 The Abrasive Wheels Regulations, 1970.

21.3 The Protection of Eyes Regulations, SI 1974 no.1681. London, HMSO, 1974.

21.4 BS 2092 : 1967 'Industrial eye-protectors'. London, British Standards Institution.

21.5 BS 1870 'Safety footwear'. London, British Standards Institution.

21.6 BS 4676 : 1971 'Gaiters and footwear for protection against burns and impact risk in foundries'. London, British Standards Institution.

21.7 BS 5304 : 1975 'Code of practice for safeguarding of machinery'. London, British Standards Institution.

Chapter 22 — Measurement and assessment of fume

22.1 BS DD54 'Methods for the sampling and analysis of fume from welding and allied processes'. London, British Standards Institution.

22.2 HSE. 'Guidance note (environmental health) Threshold Limit Values', EH15/79. London, HMSO.

22.3 OAKLEY, P.J. 'On-site measurement of welding fume'. *Welding Inst. Research Bull.*, **19** (10), 1978, 287-90.

22.4 MORETON, J. and FALLA, N.A.R. 'Analysis of airborne pollutants in working atmospheres: the welding and surface coating industries', Analytical Sciences Monograph no.7. London, The Chemical Society, 1980.

Chapter 23 — Sources of fume

23.1 BS 135, 458, 805 : 1977 'Specification for benzene, xylenes, and toluenes'. London, British Standards Institution.

23.2 BS 4487 : 1969 'Inhibited 1.1.1-trichloroethane (methylchloroform)'. London, British Standards Institution.

23.3 MORETON, J. and FALLA, N.A.R. 'Analysis of airborne pollutants in working atmospheres: the welding and surface coating industries', Analytical Sciences Monograph no.7. London, The Chemical Society, 1980.

Chapter 24 — Ventilation and fume protection

24.1　BS CP352 : 1958 'Mechanical ventilation and air conditioning in buildings'. London, British Standards Institution.
24.2　ANSI Z49.1-1973 'Safety in welding and cutting'. New York, American National Standards Institute, 1973.
24.3　BS 4275 : 1974 'Recommendations for the selection, use, and maintenance of respiratory protective equipment'. London, British Standards Institution.
24.4　BS 4667 'Breathing apparatus'. London, British Standards Institution.
24.5　'The facts about fume'. Abington, Welding Inst., 1976, 32*pp*.
24.6　JENKINS, N. and STEVENS, S.M. 'Welding fume and local fume extraction'. *Welding Inst. Research Bull.,* **21** (6), 1980, 160-64.

Chapter 25 — Lighting

25.1　IES code for interior lighting, 1977.
25.2　IES lighting guide for building and civil engineering sites.
25.3　BS 5345 'Code of practice for the selection, installation, and maintenance of electrical apparatus for use in potentially explosive atmospheres (other than mining applications or explosive processing and manufacture)'. London, British Standards Institution.
25.4　IES technical report no.1 'Lighting in corrosive, flammable, and explosive situations'.
25.5　BS 5266 'Code of practice for the emergency lighting of premises. Pt 1: 1975 — Premises other than cinemas and certain other specified premises used for entertainment'. London, British Standards Institution.

Chapter 26 — Fire

26.1　BS 4547 : 1972 'Classification of fires (European Standard EN2)'. London, British Standards Institution.
26.2　BS 5423 : 1981 'Specification for portable fire extinguishers'. London, British Standards Institution.
26.3　BS 1689 : 1957 'Galvanised mild steel fire buckets'. London, British Standards Institution.
26.4　'Factories — guide to the Fire Precautions Act, 1971'. London, HMSO.
26.5　BS 6306 'Code of practice for fire extinguishing installations and equipment on premises'. London, British Standards Institution.

Chapter 27 -- First aid

27.1　'First aid manual'. The St John Ambulance Association and Brigade, St Andrew's Ambulance Association, and The British Red Cross Society.
27.2　'Occupational first aid'. The St John Ambulance Association and Brigade, St Andrew's Ambulance Association, and The British Red Cross Society.

Chapter 28 – Legislation on welding safety in the UK

28.1 'Contract welders "half to blame" for £1M fire'. London, *Financial Times*, 24 March 1979, 3.
28.2 FIFE, I. and MACHIN, A.E. 'Redgrave's health and safety in factories'. London, Butterworth and Shaw, 1976.
28.3 SAMUELS, H. 'Factory law'. London, Charles Knight, 1969.
28.4 HEALTH AND SAFETY COMMISSION. 'Health and Safety at Work etc. Act, 1974; the Act outlined', HSC 2. London, HMSO, 1975.
28.5 HEALTH AND SAFETY COMMISSION. 'Health and Safety at Work etc. Act, 1974; advice to employers', HSC 3. London, HMSO, 1975.
28.6 HEALTH AND SAFETY COMMISSION. 'Health and Safety at Work etc. Act, 1974; advice to employees', HSC 5. London, HMSO, 1975.
28.7 DEPARTMENT OF EMPLOYMENT. 'Employment Protection Act – an outline'. London, HMSO.
28.8 DEPARTMENT OF EMPLOYMENT. 'Highly flammable liquids and liquefied petroleum gases. Guide to the regulations'. London, HMSO, 1973.
28.9 EGAN, B. 'The manual of safety representation, Pt 2: Safety Inspections', NCPC New Law Guidance no.9. London, New Commercial Publishing Company, 1978.

TITLES BY SERIES

British Standards in numerical order

23.1 BS 135, 458, 805 (three standards in one: BS 805) : 1977 'Specification for benzene, xylenes, and toluenes'.
2.1 BS 349 : 1973 'Identification of contents of industrial gas containers'.
23.1 BS 135, 458, 805 (three standards in one: BS 805) : 1977 'Specification for benzene, xylenes, and toluenes'.
1.2 BS 499 : Pt 1 : 1965 'Welding, brazing, and thermal cutting glossary'.
6.3 BS 638 : 1966 'Arc welding equipment, Section 8: Electrode holders'.
6.2 BS 638 : 1966 'Arc welding equipment, Section 11: Plugs and sockets'.
6.1 BS 638 : Pt 4 : 1979 'Specification for welding cables'.
5.3
8.5
10.1 } BS 679 : 1959 'Filters for use during welding and similar industrial
11.1 operations (viewing)'.
13.1
23.1 BS 135, 458, 805 (three standards in one: BS 805) : 1977 'Specification for benzene, xylenes, and toluenes'.
2.12 BS 1389 : 1960 'Dimensions of hose connexions for welding and cutting equipment'.

8.3	BS 1542 : 1960 'Equipment for eye, face, and neck protection against radiation arising during welding and similar operations'.
5.2, 8.2	BS 1651 : 1966 'Industrial gloves'.
26.3	BS 1689 : 1957 'Galvanised mild steel fire buckets'.
2.8	BS 1780 : 1960 'Bourdon tube pressure and vacuum gauges. Appendix D — Precautions relating to gauges for use with high pressure gas; Appendix E — Precautions relating to gauges for use with oxygen and acetylene'.
2.11	BS 1845 : 1977 'Specification for filler metals for brazing'.
21.5	BS 1870 'Safety footwear'.
21.4	BS 2092 : 1967 'Industrial eye-protectors'.
5.1, 8.1	BS 2653 : 1955 'Protective clothing for welders'.
20.3	BS 3510 : 1968 'A basic symbol to denote the actual or potential presence of ionising radiation'.
24.3	BS 4275 : 1974 'Recommendations for the selection, use, and maintenance of respiratory protective equipment'.
23.2	BS 4487 : 1969 'Inhibited 1.1.1-trichloroethane (methylchloroform)'.
26.1	BS 4547 : 1972 'Classification of fires (European Standard EN2)'.
24.4	BS 4667 'Breathing apparatus'.
21.6	BS 4676 : 1971 'Gaiters and footwear for protection against burns and impact risks in foundries'.
15.2	BS 4803 : 1972 'Guide on protection of personnel against hazards from laser radiation'.
2.10	BS 5120 : 1975 'Rubber hose for gas welding and allied processes'.
25.5	BS 5266 'Code of practice for the emergency lighting of premises. Pt 1: 1975 — Premises other than cinemas and certain other specified premises used for entertainment'.
21.7	BS 5304 : 1975 'Code of practice for safeguarding of machinery'.
26.5	BS 5306 'Code of practice for fire extinguishing installations and equipment on premises'.
8.7	BS 5330 : 1976 'Method of test for estimating the risk of hearing handicap due to noise exposure'.
18.2, 25.3	BS 5345 'Code of practice for the selection, installation, and maintenance of electrical apparatus for use in potentially explosive atmospheres (other than mining applications or explosive processing and manufacture)'.
18.3	BS 5490 : 1977 'Specification for degrees of protection provided by enclosures (electrical equipment)'.
2.7	BS 5741 : 1979 'Specification for pressure regulators used in welding, cutting, and related processes'.

British Standard Code of Practice
24.1 BS CP352: 1958 'Mechanical ventilation and air conditioning in buildings'.

British Standard Drafts for Development
26.2 BS DD48 : 1976 'Identification of fire extinguishers'.
22.1 BS DD54 'Methods for the sampling and analysis of fume from welding and allied processes'.

GOVERNMENT PUBLICATIONS

UK legislation referred to (other than in Chapter 28)
18.1 The Blasting (Castings and Other Articles) Special Regulations, 1949.
14.3 ⎫
20.1 ⎭ The Ionising Radiations (Sealed Sources) Regulations, SI 1969 no.808. 1969.
21.2 The Abrasive Wheels Regulations, 1970.
2.3 ⎫
16.3 ⎭ The Highly Flammable Liquids and Liquefied Petroleum Gases Regulations, SI 1972 no.917. 1972.
15.1 ⎫
21.3 ⎭ The Protection of Eyes Regulations, SI 1974 no.1681. 1974.
3.2 The Explosives Acts, 1875 and 1923.
3.3 SR and O no.992 The Petroleum (Carbide of Calcium) Order, 1929, amended by no.1442, 1947.

Official guides to UK legislation
       The Health and Safety at Work Act, 1974
28.4 HEALTH AND SAFETY COMMISSION. 'Health and Safety at Work etc. Act, 1974; the Act outlined', HSC 2. 1975.
28.5 HEALTH AND SAFETY COMMISSION. 'Health and Safety at Work etc. Act, 1974; advice to employers', HSC 3. 1975.
28.6 HEALTH AND SAFETY COMMISSION. 'Health and Safety at Work etc. Act, 1974; advice to employees', HSC 5. 1975.
26.4 'Factories — guide to the Fire Precautions Act, 1971'.
28.8 DEPARTMENT OF EMPLOYMENT. 'Highly flammable liquids and liquefied petroleum gases. Guide to the regulations'. 1973.
28.7 DEPARTMENT OF EMPLOYMENT. 'Employment Protection Act — an outline'.

Health and Safety at Work booklets
These are useful, inexpensive, authoritative how-to-do-it guides.

Ref. No.
      4   'Safety in the use of abrasive wheels'.

	18	'Industrial dermatitis: precautionary measures'.
	22	'Dust explosions in factories'.
	25	'Noise and the worker'.
	27	'Precautions in the use of nitrate salt baths'.
2.5	30	'The storage of liquefied petroleum gas at factories'.
9.2	32	'Repair of drums and small tanks: explosion and fire risk'.
	35	'Basic rules for safety and health at work'.
	36	'First aid in factories'.
	38	'Electric arc welding'.
	44	'Asbestos: health precautions in industry'.
2.13 4.2	50	'Welding and flame cutting using compressed gases'.

General official information

2.2 'Code of practice for the use of acetylene at pressures between 0.6 and 1.5 bar' (may become regulations in due course).

4.1 'Fires and explosions due to the misuse of oxygen'.

8.8 'Code of practice for reducing the exposure of employed persons to noise'.

14.1 'Limit switches and interlocks' in 'Electrical accidents and their causes', 30-47.

14.2 'Failure to safety of electrical control gear' in 'Electrical accidents and their causes', 26-52.

16.1 'Precautions in the use of nitrate salt baths', Form 848.

16.2 'Cautionary notice: nitrate salt baths', Form 849.

21.1 'Safety in pressure testing', General Series GS/4.

22.2 'Guidance note (environmental health) Threshold Limit Values', EH15/79.

9.4 'The safe cleaning, repair, and demolition of large tanks for storing flammable liquids', Technical Data Note 18.

9.5 'Entry into confined spaces: hazards and precautions', Technical Data Note 47.

9.6 'Hot work: welding and cutting on plant containing flammable materials'. Pamphlet HS(G) Series no.5.

PUBLICATIONS OBTAINABLE FROM THE WELDING INSTITUTE

Books

1.1 HOULDCROFT, P.T. 'Welding process technology'. Cambridge University Press, 1977.

24.5 'The facts about fume'. Welding Institute, 1976, 32*pp*.

Articles
7.1 BALCHIN, N.C. 'Electrical safety in welding'. *Welding Inst. Research Bull.*, **18** (3), 1977, 73-8; (4), 1977, 99-104; (9), 1977, 240.
22.3 OAKLEY, P.J. 'On-site measurement of welding fume'. *Welding Inst. Research Bull.*, **19** (10), 1978, 287-90.
24.6 JENKINS, N. and STEVENS, S.M. 'Welding fume and local fume extraction'. *Welding Inst. Research Bull.*, **21** (6), 1980, 160-64.

Wall charts
'Safe working with arc welding' (Ref.C5).
'Safe working with gas cutting and welding' (Ref.C6).

Films
'Facts of fume', 17min (16mm colour, optical sound). Distributors: Guild Sound and Vision, Woodston House, Oundle Road, Peterborough PE2 9PZ.

Overhead projector transparencies
'Safe working with welding and cutting' (Set A).

Miscellaneous services
Weldasearch information services
Buyers Guide — published annually and includes a section on safety equipment
Training courses are available in TWI's School of Welding Technology (SWT) and School of Applied Non-destructive Testing (SANDT)

OTHER TEACHING AND INSTRUCTIONAL MATERIAL
Film   'Weld in safety', 22min (16mm colour, optical sound). Sponsored by BOC Limited; distributors: Guild Sound and Vision, Woodston House, Oundle Road, Peterborough PE2 9PZ.
Booklets   'Welding safety' (Heating and Ventilating Joint Safety Committee, 1978).
'Safety training — welding in motor vehicle repairs' (Road Transport Industry Training Board, 1972).

ACCESS TO INFORMATION
The information sources listed above may be consulted in a number of ways; a suggested order of preference is:

(a)   Company technical library
(b)   a major central public library
(c)   a local technical college library
(d)   both Research and Professional members of The Welding Institute may borrow from its Library

The Welding Institute will be pleased to post catalogues of its publications, training aids, and films free of charge. Enquiries may be addressed to the Publications Department.

Copies of commercially published books may be ordered through almost any bookshop; a number of bookshops are agents for HMSO publications, but many specialist publications of organisations listed in Appendix 2 are available only direct.

Librarians should be able to locate and obtain loan/retention copies of any of the items listed above, even those which are currently out of print.

APPENDIX 2
USEFUL ADDRESSES

PART 1 – GENERAL
The Welding Institute
Abington Hall
Abington
Cambridge   CB1 6AL

Telephone: Cambridge (0223) 891162 (International Code + 44223 891162)
Telex: 81183 Weldex G

Others in alphabetical order
British Acetylene Association, *see* British Compressed Gases Association

British Compressed Gases Association
3 St James's Square
London   SW1Y 4JU

British Electrical and Allied Manufacturers Association
8 Leicester Street
London   WC2H 7BN

British Red Cross Society
9 Grosvenor Crescent
London   SW1X 7EJ

British Standards Institution
101 Pentonville Road
London   N1 9ND

Chartered Institute of Building Services
(Illuminating Engineering Society/Institution of Heating and Ventilating Engineers)
49 Cadogan Square
London   SW1X 0JB

Engineering Industry Training Board
54 Clarendon Road
Watford   WD1 1LB

Fire Protection Association
Aldermary House
Queen Street
London   EC4

Health and Safety Executive
Baynards House
1 Chepstow Place
Westbourne Grove
London   W2 4TF

Heating and Ventilating Joint Safety Committee
ESCA House
34 Palace Court
Bayswater
London   W2 4JG

Her Majesty's Stationery Office
49 High Holborn
London   WC1V 6HB

Illuminating Engineering Society, *see* Chartered Institute of Building Services

Industrial Safety (Protective Equipment) Manufacturers Association
69 Cannon Street
London   EC4N 5AB

Liquefied Petroleum Gas Industry Technical Association
PO Box 5
Shepperton
Middlesex

National Radiological Protection Board
Harwell
Didcot
Berkshire

Road Transport Industry Training Board
Capitol House
Empire Way
Wembley
Middlesex   HA9 0NG

St Andrews's Ambulance Association
Milton Street
Glasgow   G4 0HR

St John Ambulance Association
1 Grosvenor Crescent
London   SW1

Welding Manufacturers Association
Leicester House
8 Leicester Street
London   WC2H 7BN

Directories
These may be found useful to locate organisations which have moved from the addresses above, or their equivalents outside the UK (telephone numbers may be obtained from Directory Enquiries or are also given in these directories):

Directory of British Associations
(CBD Research Limited, Beckenham, Kent, England)

Directory of European Associations
(CBD Research Limited, Beckenham, Kent, England)

National (US) Directory of Addresses and Telephone Numbers
(Stanley R. Greenfield, Nicholas Publishing Company Inc./Bantam Books, Toronto/New York/London)

PART 2
The following organisations offer a welding fume measurement service at the time of writing (May 1980). There is also a list in The Welding Institute's annual Buyers Guide on a paid entry basis:

Chemical Laboratory
The Welding Institute
(for address see Part 1 of this Appendix)

Fume Control Group
BOC Limited
Engineering Division
Hammersmith House
London   W6 9DX   (and regional offices)

Director of Occupational Health Service
British Shipbuilders (North-East) Training and Safety Company Limited
Hebburn
Tyne and Wear

Laboratory
Foster Wheeler Power Products Limited
Brenda Road
Hartlepool
Cleveland   TS25 2BU

Chemistry Section
Electrical Metals Division
GEC Power Engineering Limited
Trafford Park
Manchester   M17 1PR

Engineering Metallurgy Department
Head Wrightson Teesdale Limited
PO Box 10
Stockton-on-Tees
Cleveland   TS17 6AZ

ICLS Laboratories Limited
288 Windsor Street
Birmingham   B7 4DW

Head of Hygiene Section
Institute of Occupational Medicine
Roxburgh Place
Edinburgh   EH8 9SU

Environmental Group
International Research and Development Company Limited
Fossway
Newcastle upon Tyne   NE6 2YD

National Occupational Hygiene Services Limited
12 Brook Road
Fallowfield
Manchester   M14 6UH

Coordinator, Information and Advisory Service
TUC Centenary Institute of Occupational Health
London School of Hygiene and Tropical Medicine
Keppel Street
London   WC1E 7HT

Occupational Health and Hygiene Unit
Department of Community Medicine
Welsh National School of Medicine
Heath Park
Cardiff   CF4 4XN

Wolfson Institute of Occupational Health
(For the attention of Mr G.E.T. Gillanders)
Level 5
University Medical School
Ninewells
Dundee

APPENDIX 3
METRIC UNITS

Unfamiliarity with the metric system is potentially dangerous. This Appendix gives approximate equivalents for some measurements appropriate to safety.

Length
1m = 1.09yd = 3.28ft = 3ft 3.7in.
1cm = 0.39in.
1mm = 0.039in.
1ft = 0.305m
1yd = 0.914m
1in. = 25.4mm

Volume
$1m^3 = 1.31yd^3$
1 litre = $0.035ft^3$
       = $61in^3$
       = 0.22gal
$1ft^3 = 0.028m^3$ = 28 litre

Velocity
1m/sec = 3.28ft/sec
1km/hr = 0.62 mile/hr
1ft/sec = 0.3m/sec
1 mile/hr = 1.61km/hr
          = 0.447m/sec

Gas flow rate
$1m^3/hr = 35.3ft^3/hr$
1 litre/hr = $0.035ft^3/hr$
1 litre/min = $0.035ft^3/min$
            = $2.1ft^3/hr$
$1ft^3/hr = 0.028m^3/hr$
          = 28 litre/hr
          = 0.47 litre/min

Mass, weight
1 kg = 2.2 lb
1 metric ton (1 tonne) = 1000 kg = 2200 lb
1 N = 0.225 lbf (pounds force, the weight of a mass of 1 lb)
1 lb = 0.454 kg
1 British ton = 1.016 metric ton = 1016 kg
1 lbf (pounds force) = 4.45 N

Pressure
1 bar = $10^5 N/m^2$ = $10^5 Pa$ (pascal)
  = 0.1 MPa
  = 14.5 lb/in^2
  = 1.02 kgf/cm^2
  = 1.02 kp/cm^2
1 lb/in^2 = 0.0689 bar
  = 0.070 kgf/cm^2
  = 0.070 kp/cm^2

Gas cylinder data

Permanent gases (argon, nitrogen, hydrogen, oxygen):
206 bar = 3000 lb/in^2
180 bar = 2500 lb/in^2

Carbon dioxide:
30-60 bar = 435-870 lb/in^2
50 lb = 23 kg
28 lb = 13 kg
400 ft^3 = 11.3 m^3
224 ft^3 = 6.3 m^3

Acetylene:
15.5 bar = 225 lb/in^2

Propane:
8 bar = 120 lb/in^2

BS 5741 welding gas pressure regulators:
Maximum inlet pressure: oxygen and other compressed gases 200 bar
    (2900 lb/in^2)
    acetylene 20 bar (290 lb/in^2)

Illumination
1 lux = 0.093ft candela
1ft candela = 1 lumen/ft^2 = 10.8 lux

Radiation
Absorbed dose: 1 gray (1Gy) = 100rad
Dose equivalent: 1 sievert (1Sv) = 100rem
Dose limits, exposed workers: 50mSv/year = 5rem/year
                                      7.5μSv/hr = 0.75mrem/hr
                general public: 5mSv/year = 0.5rem/year
Activity: 3.7 x 10^{10} becquerel (3.7 x 10^{10}Bq) = 1 curie
Exposure: 1 rontgen = 1R = 2.58 x 10^{-4} coulomb/kg

APPENDIX 4
UPDATING

Since the manuscript of this book was prepared there have been a number of changes in the field, and to make it as up-to-date as possible, these have been listed below as at January 1982.

CHAPTER 2 — Additional information
2.15 K. Harper 'Flame arrestors and associated devices — a practical safety assessment'.
Metal Construction, vol.13, no.3. March 1981, p.163.
2.15 HSE Guidance note CS4 (Chemical Safety)
The keeping of LPG in cylinders and similar containers.
2.16 HSE Guidance note CS5
The storage of LPG at fixed installations
2.17 HSE Guidance note CS6
The storage and use of LPG on construction sites.

CHAPTERS 7 and 12 — Heart pacemakers
Heart pacemakers, alternatively known as cardiac pacemakers, are used to provide effective treatment for persons suffering from illness in which the heart muscle is not triggered normally to give a stable heartbeat. The pacemakers can be affected if persons using them are in strong magnetic fields, such as those near welding equipment for arc or resistance welding. The pacemakers are implanted in the body and connected to the heart muscle, supplying regular trigger pulses unless normal muscle activity is detected by a high-sensitivity amplifier. Any induced voltages which fluctuate or pulse around once per second may simulate normal activity and inhibit pacemaker action, whereupon the patient's heart may stop or slow down, causing him to faint: on removal or cessation of the field, normal functioning would most likely be restored, however.

Large pulsed currents and magnetic fields are common in arc welding, particularly where the arc is proving hard to start in manual metal arc or gas-shielded processes, and in resistance seam welding: there does not appear to be at the time of writing (May 1981) any documented case of ill-effects from this kind of interaction. Normal walls or partitions will not screen off the low-frequency magnetic fields most likely to cause adverse effects. It would appear that the extent of any problems would be dependent on many factors, such as design of the pacemaker (there are a number of commercial products), the routing of the leads within the body, and the patient/electrode interface: even makers disagree on, for example, whether patients using their products can safely use an electric razor. Until more precise recommendations can be made, the following guidelines are suggested.

1. Advise known pacemaker users to keep away from arc and resistance welding where this is compatible with their job — or —
2. Where a pacemaker user wishes to work near arc (3m) or resistance (10m) welding equipment, or to use it himself, he should ask his general practitioner or hospital consultant to advise on the desirability of a 'trial run' under medical supervision in his intended workplace.

CHAPTER 12 — Additional information
12.1 BS 5924:1980 — Safety requirements for electrical equipment of machines for resistance welding and allied processes.

CHAPTER 16 — Additional information
16.4 'Guidelines for safety in heat treatment: Part1: Use of Molten Salt Baths' 1981, Wolfson Heat Treatment Centre, The University of Aston in Birmingham, Gosta Green, Birmingham B4 7ET.

CHAPTERS 22, 23 and 24 — Additional information
23.4 Welding fume = sources, characteristics, control. Following the compilation of a 10 year sponsored project, the Institute has published a major work of reference covering sampling, classification of consumables, particle size in welding fume, fume analysis, chromium, ozone, effect of operational factors, typical pollutants, effect of parent metal, welder protection, control of fume, general environmental effects and biological effects.

This two volume work is published by The Welding Institute (1981).

CHAPTER 24 — Personal fume protection (dust respirator)
24.6 Disposable respirators are specified in BS 6016:1980 Filtering facepiece dust respirators.

CHAPTER 28 — The notification of accidents and dangerous occurrences regulations 1980
Statutory Instrument no.804
These regulations, made under the Health and Safety at Work etc., Act 1974,

require a notifiable accident resulting in the death of or major injury to any person or a notifiable dangerous occurrence to be notified immediately to the enforcing authority (normally HSE) under the Act. Schedule 1 lists notifiable dangerous occurrences including: collapse or overturning of any lift, hoist, crane or mobile powered access platform which might have been liable to cause a major injury; explosion or fire in process materials stopping normal work for more than 24hr; when inhalation of any substance, or lack of oxygen causes acute ill health requiring medical treatment, and any unintentional ignition or explosion of explosives (unless notifiable under the Explosives Act 1875).

CHAPTER 28 — The control of lead at work regulations 1980
Statutory Instrument no.1248

These regulations, made under the Health and Safety at Work etc. Act 1974, require that where any work may expose persons to lead, the work shall be assessed to determine the nature and degree of the exposure: records must be kept of the assessment and any further required actions. Control measures must so far as reasonably practicable be effective without the use of personal respiratory protective equipment: where such equipment is required, it must be of a type approved in writing by HSE.

Medical surveillance is required, where an individual's exposure is significant. Employees must be given training as needed to ensure their own safety, to ensure that they can effectively carry out assessments, in the proper use and maintenance of protective equipment, etc.

'Control of Lead at Work — Approved Code of Practice' (HSE/HMSO 1980) gives practical guidance. Lead burning, welding and cutting of lead coated and painted work, and dry grinding of lead are listed as types of work where there is liable to be significant exposure to lead; soldering and handling of clean solid metallic lead are listed as those where there is not liable to be significant exposure. Measurements will be required where there is liable to be significant exposure, unless there is sufficient suitable written information available, for example from suppliers of consumables, for an adequate assessment. Appendix 1 gives a lead-in-air standard of $0.15mg/m^3$ (as Pb, 8hr time-weighted average concentration): exposure to levels half this or less is not considered 'significant', therefore medical supervision is not required. The employer should distribute the leaflet 'Lead and You' to each employee.

CHAPTER 28 — The health and safety (first-aid) regulations 1981

These regulations, made under the Health and Safety at Work etc. Act 1974, are due to come into force on 1 July 1982. They repeal the First-Aid Boxes in Factories Order 1959, and the First-Aid (Standard of Training) Order 1960, which were listed in Chapter 28. Practical guidance is given in the Health and Safety series booklet HS(R) 11 'First Aid at Work', which contains the text of the regulations, guidance notes, and the approved code of practice.

The employer is required to make provision, as appropriate to the circumstances, of equipment, a first-aid room, and first-aiders, and to inform his employees of the arrangements he has made: self-employed persons must provide appropriate equipment. The circumstances to be taken into account are spelt out in the Approved Code of Practice — for example special or unusual hazards, numbers and distribution of employees, the location of the establishment and work by employees on their own or in small groups.

APPENDIX 1 — Section (C) — Titles by series
Health and Safety at Work: Booklet 38, Electric Arc Welding (HSE/HMSO). Currently (November 1981) out of print, undergoing major revision, and not expected to be available until 1982.

INDEX

Abrasions, 157
Abrasive Wheels Reg, 164
A.C., 47
Acetone, 119
A.C.H., 134
Acetylene, 17, 19, 102
Acetylene generators, 28
Acetylene order, 165
Acids, 89, 90
Addresses, 181
Air-arc cutting, 7, 58
Air changes per hour, 134
Air-cooled transformers, 41
Air-fed welding helmet, 139
Air supply, 138
Alkalis, 90
Alternating current, 47
Aluminium, 126, 130
Aluminium alloys, 122
Aluminium bronze, 122
Aluminium dust, 101
Ammonia, 90
Apron, 53
Arc-air cutting, 7
Arc cutting, 58
Arc eye, 156
Arc stud welding, 7
Arc welding equipment, 40
Argon, 131
Argon arc welding, 4
Artificial respiration, 49
Asbestos, 37, 120
Associated workers, 58
Atomic hydrogen welding, 8
Automatic metal arc welding, 6, 130

Back-pressure valves, 30
Backfeeding, 25
Backfire, 26, 33
Barrier cream, 89
Basic electrodes, 129
Becquerel, 108
Benzene, 89, 119
Beryllium, 79, 87, 121, 129
Bibliography, 168
Blasting, 97

Blasting Regulations, 163
Blowpipes, 24
Blowpipes, low pressure, 30
Boiling, 62
Borax, 86
Boron trifluoride, 88
Brass, 122
Braze welding, 1, 2, 84
Brazing, 1, 84
Breathing apparatus, 63, 139
British Standards, 175
Bulk storage of gases, 23
Burns, 34, 49, 70, 157
Butt welding, 8

Cable connectors, 42
Cable, welding, 42
Cadmium, 71, 86, 99, 121, 124, 129, 160
Caesium, 108
Calcium carbide, 28, 31
Calcium fluoride, 129
Cap, welder's, 53
Carbide of Calcium order, 166
Carbon arc welding, 4
Carbon dioxide, 131
Carbon dioxide laser, 81
Carbon monoxide, 127, 131, 132, 159
Carbon tetrachloride, 89, 119
Cardboard, 146
Cardiac pacemakers, 70, 189
Cataract, 82
Caustic soda, 120
Cellulosic electrodes, 129
Chains, 49
Chlorinated hydrocarbons, 119
Chromium, 124
Cigarette lighters, 146
Circuits, welding, 42
Clamps, earthing, 42
CO_2 welding, 5, 44, 49, 50
Cobalt, 108
Cold pressure welding, 12
Combustibles, 60, 145
Competent person, 111
Compressed gases order, 166
Connectors, cable, 42

Constricted-arc welding, 7
Construction Reg., 164
Consumable-guide welding, 10, 68
Contact lens, 57
Control of Lead Reg., 191
Controlled hydrogen electrodes, 129
Copper, 121, 124, 128, 129
Crucible, 72
Crush injury, 70
Cupronickel, 122
Curie, 108
Curtains, 51
Cutting, 114
Cutting, arc, 58
Cyanide, 88
Cylinders, gas, 18
Cylinder stores, 22

D.C., 47
Degreasing, 97
Degreasing tanks, 119
Dermatitis, 94
Detection, fire, 147
Dielectric welding, 11
Dip brazing, 14, 85
Dip soldering, 14
Direct arc, 65
Direct current, 47
Discharge of polluted air, 137
Disposable respirator, 190
Distance, safe, 52
Dosemeter, 109
Double insulation, 49, 95, 114
Dross, solder, 94
Dust explosion, 74, 79
Dust respirator, 138, 190
Dust, 115, 146

E.R.W., 8
Ear muffs, 59
Ear plugs, 59
Ear protection, 58, 66, 71
Earth lead, 114
Earthing, 40, 48
Earthing clamps, 42
Electric arc welding, 6
Electric heating, 49
Electric resistance welding, 8
Electric shock, 47, 49, 76, 82, 90, 95, 153
Electricity Regulations, 163
Electrode holders, 42
Electron-beam brazing, 11
Electron-beam welding, 11, 76

Electronically-switched filters, 57
Electroplating, 124
Electroslag welding, 9, 68
Electrostatic precipitator, 137
Emergency exit, 147
Emergency lighting, 144
Emergency, ionising radiation, 111
Employment Protection Act, 166
Etching, 120
Exhaust, 132
Explosion, 32, 60, 74, 86, 100, 107
Explosives Act, 165
Exposed workers, 109
Extinguisher, fire, 149
Eye injury, 34, 50, 155
Eye protection, 38, 55, 71, 75, 83, 114

Face shield, clear, 55
Factories Act, 162
Filter, electronically switched, 57
Filter, fume, 137
Filter, liquid crystal, 57
Filter, viewing, 38, 44, 55, 66, 68, 75, 99
Fire, 34, 51, 74, 96, 100, 107, 145
Fire alarm, 147
Fire brigade, 151
Fire detection, 147
Fire extinguisher, 149
Fire, gas, 33
Fireman, standby, 147
Fire Precautions Act, 162
First aid, 153
First aid boxes order, 163
First aid equipment, 160
First aid order, 164
First aid Regs., 191
Fittings, gas, 25
Fittings, low pressure gas, 30
Flame arrestors, 189
Flame cutting, 3
Flame processes, 131
Flash welding, 8, 9
Flash back, 26
Flashback arrestor, 26, 189
Fluoride, 86
Fluorspar, 129
Flux, 68, 87
Flux, solder, 94
Flux bath brazing, 85
Flux-cored arc welding, 5
Foam, plastics, 146
Footwear, 114
Fresh air supply, 138

Friction welding, 12
Fume, 35, 51, 66, 71, 86, 106, 159
Fume extractor gun, 136
Fume measurement, 115
Fume measurement service, 183
Fume protection, personal, 138
Fume sources, 118
Furnace brazing, 13, 85
Fusarc welding, 6

Galvanised metal, 35
Galvanising, 124
Gamma rays, 108
Gas cutting, 3
Gas, inert, 61
Gas leaks, 34
Gas metal arc welding, 5
Gas, pollutant, 115, 159
Gas pressure welding, 13
Gas-shielded arc welding, 44, 48, 50
Gas tungsten arc welding, 4
Gas welding, 2
Gauges, 24
German silver, 122
Glass reinforced plastics, 146
Gloves, 37, 54, 71
GMAW, 5
Goggles, 38, 71, 91
Goggles, clear, 55
Gouging, 3
Government publications, 177
Gravity welding, 6
Gray, 108
Grease, 146
Grinding, 114
GTAW, 4
Guards, 70
Guards, machinery, 114
Gunmetal, 122

Half-mask respirator, 138
Handrails, 112
Handshield, 38, 44, 54
Health and Safety at Work Act, 161
Health and Safety (First Aid) Reg., 191
Heart pacemakers, 70, 189
Heat stroke, 159
Heated tool welding, 15, 105
Heliarc, 4
Helium, 131
Helium-neon laser, 81
Heliweld, 4
Helmet, 34, 44, 54

Helmets, safety, 114
H.f., 44, 48, 50, 65, 85, 90, 106
H.f. welding, 11
High frequency, 44, 48, 50, 65, 85, 90, 106
High frequency welding, 11
Highly flammable ... Reg., 165
High pressure plant, 11
High yield steel, 122
Holders, electrode, 42
Hose, 24
Hose check valves, 25
Hot gas welding, 15, 104
Hydrofluoric acid, 88, 90, 95, 120
Hydrogen, 19, 90
Hydrogen cyanide, 88

Incoloy, 122
Inconel, 122
Induction brazing, 11, 85
Induction heating, 11
Induction soldering, 11
Induction welding, 11
Inert gas, 61
Inert gas forming, 61
Infra-red, 56, 82
Inner shield, 5
Insulation, thermal, 146
Internal combustion engines, 131
Ionising radiation, 108
Ionising radiation warning sign, 109
Ionising radiation Reg., 164
Iridium, 108
Iron oxide, 126
Iron powder electrodes, 129
Irrespirable atmosphere, 139

Keyhole, 66

Lacerations, 70
Lagging, asbestos, 120
Lancing, 3
Laser cutting, 16, 81
Laser welding, 16, 81
Lead, 93, 99, 121, 124, 129
Lead-coated metal, 35
Lead Regulations, 191
Leaks, gas, 34
Legislation, 161, 190
Lens, contact, 57
Leukaemia, 108
Lifting, 112
Lighting, 143
Lighting, emergency, 144

195

Liquid crystal filter, 57
Liquid petroleum gas, 189
Local extraction, 135
Lost wax, 73
Low hydrogen electrodes, 129
Low pressure plant, 28
Low voltage safety device, 47
L.P.G., 189
L.P.G. Regs., 165
L.V.S.D., 47

Macroetching, 120
Macrosection, 120
MAG welding, 5
Magnesium, 145
Magnesium alloys, 122
Major wounds, 157
Manganese, 121
Manganese bronze, 122
Manganese steels, 121
Manifolding gas cylinders, 20
Manipulators, 113
Manual lifting, 112
Manual metal arc welding, 6, 129
Mechanical hazards, 112
Mechanical testing, 114
Medium pressure, 28
Metal active gas welding, 5
Metal arc gas shielded welding, 5, 44, 49, 50, 130
Metal fume, 159
Metal fume fever, 98, 124
Metal inert gas welding, 5
Metal spraying, 2, 97
Metric units, 186
Microplasma welding, 7, 65
MAG welding, 5
MIG welding, 5, 44, 49, 50, 130
Mild steel, 121, 122
Minor cuts, 157
MMA welding, 6, 129
MMA, multiple-operator, 48
Moisture, 72
Monel, 122
Mould, 73
Multiple-operator MMA, 48

Naptha, solvent, 119
Natural gas, 90, 101
Needle-arc welding, 7
Neodymium laser, 81
Neon, 81
Nickel alloy coatings, 122

Nickel, 124
Nimonic, 122
Nitric oxide, 128
Nitrile rubber, 102
Nitrogen, 131
Nitrogen oxides, 159
Noise, 71, 99
No-load, low voltage device, 47
Non-return valves, 25
Non-transferred arc, 65
Notification of Accidents Reg., 190

O.C.V., 47, 65
O.C.V. reduction relay, 47
Oil, 71, 146
Oil-filled transformers, 40
Oil mist, 78
Open circuit voltage, 47, 65
Overall, 53
Overhead cranes, 49
Oxyarc cutting, 6
Oxygas cutting, 3
Oxygas flames, 131
Oxygen, 18, 101
Oxygen arc cutting, 6
Oxygen cutting, 3
Oxygen enrichment, 34, 146
Ozone, 66, 127, 138, 159

Pacemakers, heart, 70, 189
Painting arc welding booths, 50
Paints, 125, 126
Penetrant testing, 120
Perchloroethylene, 89
Permit to work, 63, 140
Personal fume protection, 138
Petroleum Act, 166
Phosgene, 159
Phosphor bronze, 122
Pickling, 89
Plasma arc, 65
Plasma cutting, 7
Plasma spraying, 7
Plasma welding, 7
Plastics, 99, 103, 125
Plastics coating, 126
Plastics foam, 146
Plastics spraying 14, 107
Plating, 124
Plutonium, 79
Pollutant constituents, 123
Polystyrene foam, 127
Polytetrafluorethylene, 107

Polyurethane foam, 127
Portable electric tools, 49
Portable generators, 30
Positive pressure respirator, 139
Powder cutting, 3, 35
Precipitator, electrostatic, 137
Pressure testing, 113
Projection welding, 8, 69
Propane, 19, 101
Props, 112
Protection, ear, 58
Protection of eyes Regs., 165
Protective clothing 36, 53
PTFE, 107

Rad, 108
Radiation, ionising, 77
Radiation, ionising Regs, 164
Radiation safety officer, 111
Radiofrequency welding, 11
Radiography, 108
Regulators, 23
Rem, 108
Rescue plan, 147
Rescue, fire, 147
Resins, 99
Resistance brazing, 10, 85
Resistance soldering, 10
Resistance welding, 8, 69
Respirator, 138
Respirator, disposable, 190
Retina, 82
Rings, 90
RF welding, 11
Rosin, 94
Rotary electrical equipment, 41
Ruby laser, 81
Rutile electrodes, 129

Safe distance, 52
Safe platform, 112
Safety belt, 112
Safety footwear, 114
Safety helmets, 114
Safety spectacles, 55
Salt bath brazing, 14, 85
Salt baths, 190
Screens, arc, welding, 51
Seam welding, 8, 69
Security, 146
Self-shielded arc welding, 5
Semi-automatic welding, 5
Shielded metal arc welding, 6

Shielding gas, 21
Shipbuilding Regs, 164
Shock, electric, 47, 49, 76, 82, 90, 95, 153
Shock, traumatic, 158
Siderosis, 130
Sievert, 108
Sleeves, 53
Slings, 49
SMAW, 6
Smoke, 115
Smoking, 146
Soft solder, 122
Solder, 122
Soldering, 1, 93
Solvent cleaning, 118
Solvent naptha, 119
Solvents, 146
Spats, 53
Spatter, salt, 91
Spitting, metal, 96
Spot welding, 8, 69
Spraying, 65
Spraying, plastics, 107
Sprinkler, fire, 147
Stainless steel, 122
Stainless steel, flux, 95
Standby fireman, 147
Stationary generators, 28
Steaming, 62
Stick welding, 6
Storage, radioactive sources, 110
Stud welding, 7
Submerged-arc welding, 5, 130
Surface treatment, 118
Surfacing, 1

Tack welds, 112
Thermal cutting, 2
Thermal spraying, 14, 92
Thermit welding, 11, 12, 72
Thoriated electrode, 131
Threshold limit values, 116
TIG welding, 4, 44, 48, 50, 130
Titanium dioxide, 129
Toeboard, 112
Toluene, 119
Torch brazing 2, 84
Torch soldering, 2
Touch welding, 6
Toxic atmosphere, 63
Transferred arc, 65
Transformers, aircooled, 41
Transformers, oilfilled, 40

Traumatic shock, 158
Trichloroethane, 119
Trichloroethylene, 89, 118
Tungsten arc gas shielded welding, 4, 44, 48, 50, 130
Tungsten inert gas welding, 4, 44, 48, 50, 130

Ultraviolet light 56, 120

Ventilation 100, 133, 140
Vessels, precautions for 60
Viewing filter, 68, 75
Visible light, 82

Water spray, 67
Water tank, 67
Water tray, 124

Waxes, 99
Welder position, 134
Welder's lung, 130
Welding cables, 42
Welding circuits, 42
Welding Institute publications, 178
Welding, definition, 1
White spirit, 119
Wire feed units, 113
Wire ropes, 112
Wood, 145
Workers, associated, 58
Work positioners, 112

X-rays, 77, 108

Zinc, 71, 87, 121, 124, 126
Zinc chloride, 94
Zinc chromate, 126